FOOD AND BEVERAGE CONSUMPTION AND HEALTH

CONSUMPTION AND CONTAMINATION OF DAIRY PRODUCTS

FOOD AND BEVERAGE CONSUMPTION AND HEALTH

Additional books and e-books in this series can be found on Nova's website under the Series tab.

FOOD AND BEVERAGE CONSUMPTION AND HEALTH

CONSUMPTION AND CONTAMINATION OF DAIRY PRODUCTS

EGOR VAGIN
EDITOR

Library of Congress Cataloging-in-Publication Data

Names: Vagin, Egor, editor. Title: Consumption and contamination of dairy products / Egor Vagin, editor.
Description: New York : Nova Science Publishers, [2020] | Series: Food and beverage consumption and health | Includes bibliographical references and index. | Summary: "In Study about the Consumption and Contamination of Dairy Products in Two Countries, the authors first describe the microbiological, physicochemical and sensory properties of a traditional Michouna cheese produced from goat and cows' milk in East of Algeria. A description of the microbiological analyses of 128 samples of different dairy products produced by an establishment located in the Marche region, Central Italy, is provided. The results demonstrate good manufacturing conditions. One of the main fermented milk drinks consumed in Algeria, lben, is characterized, recording the absence of all pathogenic micro-organisms and mold. To examine the consumers' acceptance of bakery products incorporating whey residue, a by-product of the cheese industry, a descriptive cross-sectional study was conducted through a questionnaire survey on a non-probabilistic sample composed of 299 participants. In closing, the authors present some results of a questionnaire survey carried out in Portugal and Brazil, investigating the consumption habits of some classes of dairy product"-- Provided by publisher.
Identifiers: LCCN 2020041239 (print) | LCCN 2020041240 (ebook) | ISBN 9781536186543 (hardcover) | ISBN 9781536186741 (adobe pdf) Subjects: LCSH: Dairy microbiology. | Dairy products. | Dairy products--Contamination. | Dairy products industry.
Classification: LCC QR121 .C93 2020 (print) | LCC QR121 (ebook) | DDC 613.201/579--dc23
LC record available at https://lccn.loc.gov/2020041239
LC ebook record available at https://lccn.loc.gov/2020041240

Published by Nova Science Publishers, Inc. † New York

CONTENTS

PREFACE

In the study about the *Consumption and Contamination of Dairy Products in Two Countries*, the authors first describe the microbiological, physicochemical and sensory properties of a traditional Michouna cheese produced from goat and cows' milk in East of Algeria.

A description of the microbiological analyses of 128 samples of different dairy products produced by an establishment located in the Marche region, Central Italy, is provided. The results demonstrate good manufacturing conditions.

One of the main fermented milk drinks consumed in Algeria, lben, is characterized, recording the absence of all pathogenic micro-organisms and mold.

To examine the consumers' acceptance of bakery products incorporating whey residue, a by-product of the cheese industry, a descriptive cross-sectional study was conducted through a questionnaire survey on a non-probabilistic sample composed of 299 participants.

In closing, the authors present some results of a questionnaire survey carried out in Portugal and Brazil, investigating the consumption habits of some classes of dairy product.

Chapter 1 - This is the first study of this type of cheese and the first report describing microbiological, physicochemical and sensory properties of a traditional *Michouna* cheese produced from goat and cows' milk in East of Algeria. Three cheese samples were prepared from goat and cow milk in the laboratory according to the traditional method used by people in Tebessa areas. The average values of pH studied cheeses vary between 6.04, 5.85 and 5.90 for cow's cheese (CA), goat's cheese (CB) and the cheese mixing between cow and goat milk (CC) respectively, the total solids of *Michouna* of three manufacturing are 463,2g/Kg (CA), 472g/Kg (CB) and 461.3g/Kg (CC), with the fat content of three cheeses ranged from 175g/Kg, 194g/Kg to 205g/Kg. Proteins *Michouna* constitute the second major component with a rate of 249 g/Kg, 207g/Kg and 193g/Kg respectively for the cheese CA, CB and CC. The authors' microbiological results show anintense presence of lactic acid bacteria, the absence of pathogens and the presence of microorganisms of contamination with low rate make *Michouna* an excellent microbiological quality cheese. Rich *Michouna* in aroma is explained mainly by the wealth of raw material used. Cheese prepared from goat milk and cow *Lben* (CC) remains the most popular and most preferred. About the hedonic test it seems that CC cheese is the most preferred.

Chapter 2 - In Italy, the manufacture of dairy products is an ancient tradition that allows the placing on the market of a wide variety of fresh and ripened cheeses, strongly linked to their place of origin. Their quality and safety rely primarily on the raw matter, i.e., milk that can be contaminated by microorganisms originating from udder infections, farm environment and feedstuffs, as well as milking and processing equipment. Other sources are represented by soil, faeces and bedding material attached to teats and released into milk during its collection. The microbial contamination of raw matter is a great concern for cheesemakers, as some spoilage microorganisms can affect the sensory characteristics of the products through the enzymatic alteration of milk components. However, also during the cheese production process a potential contamination can occur, due to the lack of good hygiene practices, failure of thermal treatment operations and/or temperature abuse during the storage. Moreover, spoilage microorganisms can also derive from hygiene mistakes made by the personnel working in the dairy industries. Finally, it must be highlighted that some pathogens can be present in milk and derived products, thus becoming dangerous for consumers. The Commission Regulation (EC) No 2073/2005 and its following amendments established the microbiological criteria that should be monitored by the food business operators to ensure the food safety of their products, as well as the respect of hygiene conditions during processing. The aim of this chapter was the description of the microbiological analyses made on 128 samples of different dairy products produced by an establishment located in the Marche region, Central Italy. The results showed the compliance with the regulatory criteria, as well as the absence of pathogens, therefore demonstrating good manufacturing conditions.

Chapter 3 - The present study objective is to characterize one of the main fermented milk drinks consumed in Algeria "*lben.*" 30 *lben* samples prepared from cow milk are collected commercially and 29 samples of *lben* prepared from goat's milk are collected from 3 farms located in Tebessa. *Lben* is characterized by its acidic taste with a pH of 4.6 and relatively low dry matter which varies between 90.6 g/L and 97.2 g/L for cow and goat respectively. The fat content is low (8.95 g/L and 13.5 g/L). The hygienic quality seems satisfactory. The authors record the absence of all the pathogenic micro-organisms sought and the mold. Total aerobic mesophilic flora is presented with a relatively high rate. The authors' results show an intense presence of lactic bacteria (*Lactococci*, *Lactobacilli* and lactic *Streptococci*).

Chapter 4 - By-products and residues from dairy industry are rich sources of nutritive components. Since their elimination brings some environmental problems, finding alternatives for their utilization assumes a particular importance, both from the economic as well as environmental points of view. In view of these challenges, the purpose of the work presented here was to examine the consumers' acceptance of bakery products incorporating whey residue, a by-product of the cheese industry. This descriptive cross-sectional study was conducted by questionnaire survey on a non-probabilistic sample composed of 299 participants aged 18 and over. Results showed that 77% of the participants eat bread on a daily basis, mostly at breakfast (65.8%) and preferred freshly baked bread. Most of them opted for bread made with rye flour (53.1%). The choice of bread was influenced by age, education and the practice of balanced diet. As for cookies, they were consumed by 56.5% of the participants on a regular basis, mainly as a snack between meals, 2 or 3 times/week. They favoured simple formulations, without fillings or flavourings. The majority of the respondents seemed to have some knowledge about whey, which was influenced by age, education, following a balanced diet and practicing exercise. Around 41% of the participants admitted they would consume bakery products made with whey residues, mostly due to their additional protein content followed by concerns about the environment and variability in taste. Considering the increasing demand for Serra da Estrela cheese, and consequent increase in by-products, their utilization in food products can be helpful to reduce water and soil pollutants while stimulating local economy.

Chapter 5 - The present chapter presents some results of a questionnaire survey undertaken on a sample of participants from Portugal and Brazil, which investigates the consumption habits of some classes of dairy products. This is a descriptive cross-sectional study made on a non-probabilistic convenience sample, consisting of 850 participants, 430 from Brazil and 420 from Portugal. The results obtained showed that in both countries the consumption of milk and milk products as well as cheeses, butters or yogurts was very low. Nearly half or more than half of the participants never consumed these dairy products, and those who consumes them did it only once or 2-3 times per week. Small differences were identified in the dairy consumption patterns in both countries, with slightly more Portuguese participants consuming milk, cheese, butter and yogurt as compared with Brazilians.

In: Consumption and Contamination … ISBN: 978-1-53618-654-3
Editor: Egor Vagin

Chapter 1

COMPARATIVE STUDY OF THE MICROBIAL, PHYSICOCHEMICAL AND SENSORY QUALITY OF A TRADITIONAL CHEESE *MICHOUNA* PRODUCED FROM COW AND GOAT'S MILK

Meriem Derouiche[1,*] and Hacène Medjoudj[2]
[1]Laboratory of Food Engineering, Institute of Nutrition, Food and AgroFood Technologies (INATAA) University of Constantine 1, Constantine, Algeria University of Constantine
[2]Institute of Sciences and Applied Techniques (I.S.T.A), University Larbi Ben Mhidi of Oum El Bouaghi, Algeria

ABSTRACT

This is the first study of this type of cheese and the first report describing microbiological, physicochemical and sensory properties of a traditional *Michouna* cheese produced from goat and cows' milk in East of Algeria. Three cheese samples were prepared from goat and cow milk in the laboratory according to the traditional method used by people in Tebessa areas. The average values of pH studied cheeses vary between 6.04, 5.85 and 5.90 for cow's cheese (CA), goat's cheese (CB) and the

* Corresponding Author's E-mail: meriemderouiche@yahoo.fr.

cheese mixing between cow and goat milk (CC) respectively, the total solids of *Michouna* of three manufacturing are 463,2g/Kg (CA), 472g/Kg (CB) and 461.3g/Kg (CC), with the fat content of three cheeses ranged from 175g/Kg, 194g/Kg to 205g/Kg. Proteins *Michouna* constitute the second major component with a rate of 249 g/Kg, 207g/Kg and 193g/Kg respectively for the cheese CA, CB and CC. Our microbiological results show anintense presence of lactic acid bacteria, the absence of pathogens and the presence of microorganisms of contamination with low rate make *Michouna* an excellent microbiological quality cheese. Rich *Michouna* in aroma is explained mainly by the wealth of raw material used. Cheese prepared from goat milk and cow *Lben* (CC) remains the most popular and most preferred. About the hedonic test it seems that CC cheese is the most preferred.

Keywords: *Lben*, *Michouna*, microbiological quality, physicochemical characteristics, sensory analysis, Tebessa

1. Introduction

Although our country has a delay in the cheese industry as compared to western countries, some products have a long history in Algeria and are traditionally made by ancestral processes from cow's milk, sheep or goat. These products are very difficult reflected in official statistics. Traditional Algerian dairy products, especially the fermented types, have been the pride of culinary tradition for centuries and have played a major role in the diet of communities in the rural region. A variety of cheeses are made in the Algerian territory even if the activity is limited to the domestic scale. Ten (10) types of traditional cheese are produced in different area of Algeria, but the most known are only *Klila* and *Jben* (Hallel, 2001). Others remain unknown compared to traditional Spanish, Turkish and Moroccan products. *Michouna*, a product manufactured in the east Algeria especially in region of Tebessa located 560 Km from Algiers and 40 Km from the Algerian-Tunisian border, is an agro-pastoral area. The intense presence of *Michouna* is located in the rural cities; production is for local or for family consumption (Derouiche & Zidoune, 2015). *Michouna* is a fresh cheese frequently produced especially in rural areas. It can be prepared all year round. The frequency of the preparation of this cheese is mainly linked to the availability of the raw material. Traditionally, *Michouna* is prepared with goat's milk, but currently cow's milk is frequently used, this cheese can be prepared by mixing milk and *Lben* cow, milk and goat *Lben* or from goat milk and cow *Lben* (Derouiche et al., 2017). Rare information is available on how to make, consume and store this cheese, also data on its physicochemical, microbiological and sensory characteristics. The objective of this study is to carry out manufacturing trials for this cheese, by applying the traditional diagram from previous studies in order to test its reliability, as well as to carry out physicochemical, microbiological and sensory characterizations to make better known cheese *Michouna*.

2. Materials and Methods

2.1. Raw Material

Three fabrications were made from cow's milk and *Lben* (fermented milk) prepared from cow's milk (CA), goat's milk and goat's *Lben* (CB) and goat's milk and cow's *Lben* (CC). Raw cow's and goat's milk used for the various products is fresh milk from farms located in the Tébessa region, with herds of different species (cow, goat and sheep). Salt used for salting is the iodized salt was used (ENASEL, Algeria).

2.2. Cheese-Making Diagram

Traditional diagram of *Michouna* cheese-making is illustrated in Figure 1. The *Michouna* cheeses were prepared of each type in our laboratory according to the traditional diagram described by Derouiche and Zidoune (2015). The milk used for manufacturing is that of Charolais cow race and *Arabia* goat race. The first step is the preparation of *Lben* according to the traditional method, which begins with the fermentation of milk (cow and goat), churning the resulting fermented milk is made in the *Chekoua*, with the addition of lukewarm water, with butter extraction at end of operation (or process). Cheese making start by a heat treatment in which milk is heated to boiling. The *Lben* is then added with salt, the assembly is heated a second time to the coagulation where there will be separation of curds and whey instantly. The curd is separated from the whey by filtration first through in couscoussier (perforated metal utensil) then in a cloth (cheche or muslin) suspended and drained to allow the total elimination of whey. Cheese is recovered and preserved in glass containers in a refrigerator. The conservation of this cheese should not exceed 6 days (Figure 2).

Sampling

The analyses were carried out on cow's and goat's *Lben* (n = 5), raw cow's and goat's milk (n = 5) and *Michouna* cheeses CA, CB and CC (n = 5) made in the laboratory and 5 samples collected from farms located in the province of Tebessa.

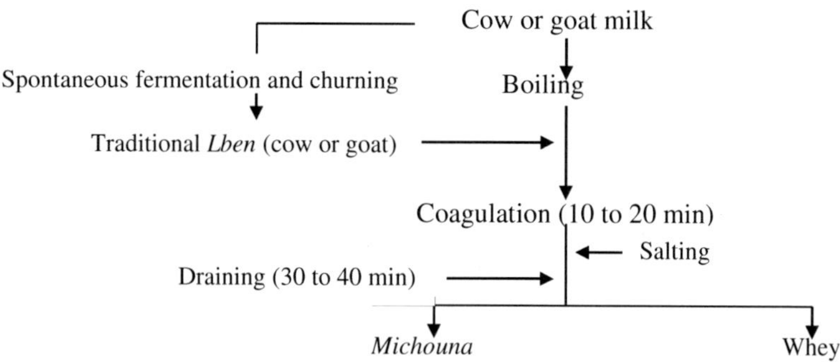

Figure 1. Traditional *Michouna* cheese making diagram (Derouiche et Zidoune, 20015).

Figure 2. *Michouna* cheese.

2.3. Physicochemical Analysis

Dry Matter (DM) was determined using a forced-air, oven-drying method at 103°C for 14 h (AFNOR, 1993) a cheese fat by the Van Gulik method (AFNOR, 1993)b, milk and *Lben* fat by Gerber method (AFNOR, 1993)c, total nitrogen (TN) and soluble nitrogen by the Kjeldahl method (FAO, 1994)**,** for the content of protein a conversion factor of 6.38 was used. Titratable acidity was measured after titration with NaOH (N/9). The pH was determined electrometrically by means of a pH meter. The Calcium is determined by flame spectrophotometry (Serres et al., 1973), the ash content was determined by the method described by FAO (1994) standard. Sodium chloride assay was performed according to the Volhart method (Audigié et al., 1984)**.** The density determination was made on the milk and whey, performed by the lacto-densimeter.

2.4. Microbiological Analysis

Twenty-five (25) g of cheese, previously heated to 45°C, was aseptically sampled and mixed in 90 ml of sterile 2% citrate buffer, pH 7.5. The cheese was blended with Ultra-Turrax (1min 9000 $r.min^{-1}$) (Guiraud, 1998). Decimal dilutions from this solution, the milk and *Lben* were made using sterile quarter-strength Ringers solution as diluents. The dilutions were used for the enumeration of bacteria by conventional microbiological methods. Enumeration of bacteria was carried out in triplicate. Mesophilic aerobic flora was counted on PCA (DIFCO) [72 h/30°C], *Lactococci* were enumerated on M17 agar (Difco), after incubation at 30°C for 3 days. *Lactobacilli* were enumerated on de Man Rogosa Sharpe (MRS) agar (Difco) after sowing in double layer and incubation at 30°C for three days; lactic streptococci were counted on M17 agar after incubation at 37°C for 2 days. Yeasts and moulds were enumerated on Oxytetracycline Glucose Agar (DIFCO) after incubation at 25°C for 5 days. Total coliforms are sought on BCPL incubated at 37°C for 24 to 48 hours; the tubes used are those considered to be positive will be the subject of the search of fecal coliforms. The search for fecal *Streptococci* is done in two steps, of first test presumption by seeding Roth incubated at 37°C for 24-48 hours and the confirmatory test will be performed in the Litsky medium and incubated for 48 hours at 37°C. The *Salmonella* requires four phases before isolation on medium SS. The *Staphylococci* were isolated after enrichment on Chapman medium and incubated at 37°C for 48 hours. After a morphological description counting different genres, the interpretation of the results includes directives of Algerian official newspaper regarding a fresh cheese, and the number of pathogenic colony tolerated in a cheese product must be zero.

2.5. Sensorial Analysis

Sensory evaluation was carried out to describe, with scoring test, the characteristics of the obtained cheese through tasting and visual observations according to Bérodier et al. (1997) a panel of fifteen untrained persons evaluated cheese texture, odor, aroma and flavor. About Ten g of cheese are placed in a dish in the laboratory and warmed at room temperature for one hour and then presented to examiners. All the characters are rated from 1 to 7. The comparison between the three *Michouna* cheeses was also performed in addition the hedonic test was performed in order to identify the preferred cheese by comparison in pairs (AFNOR, 1984).

2.6. Characteristics of Cheese Processing

These characteristics were determined by calculating certain parameters. Since dairy processing is performed in a traditional way, the use of a factor which affects a quantity of product to the amount of milk (kg or liters) necessary for producing a kg of product is interesting (spontaneous fermentation of the mass ferment not included) (Meyer & Duteurtre, 1998). The yield is calculated using the following formula:

$$\text{Yield} = (\text{Cheese mass/weight of milk}) \times 100$$

Moisture free-fat cheese (Mietton 1986):

$$\text{HFD} = 100 \text{ x } (\text{Cheese moisture/Weight of the cheese} - \text{MG})$$
$$\text{Fat/Dry} = (\text{Fat/Dry}) \text{ x } 100$$

2.7. Statistical Analysis

In this study, data were presented in the form of average ± standard deviation. ANOVA test for comparing means was used to compare if there are differences between the three cheeses.

3. Results and Discussion

Analysis results are presented in the form of average ± standard deviation of three manufacturing prepared in the laboratory (CA, CB and CC). As there is no previous work done on the *Michouna* cheese we compare the results obtained with a traditional cheeses *Klila* and *Jben*.

3.1. Physicochemical Characteristics of Raw Materials

The results of the physicochemical analysis of raw materials are presented in the following Tables 1 and 2:

The variability of the pH and the titratable acidity of raw milk are linked to hygienic conditions during milking, to the total microbial flora, to its metabolic activity as well as to the handling of milk, the casein, salt content, minerals and ions (Alais, 1984; Mathieu, 1998). The density of goat milk is comparable to that of cow's milk with an average of 1.032 and 1.030 respectively.

The results show that goat milk is significantly richer in total dry extract (194.6 g/L) and fat (52 g/L), while the protein content is similar for cow's milk and goat's milk 33.2 g/Kg and 33.4 g/L respectively.

The composition of milk in fat and nitrogenous matter is under the influence of many factors: the animal (species and breed), the environment, the lactation stage, hormonal secretions, age, temperature and duration daily illumination (Favier, 1985; Hoden & Coulon, 1991), subsequently it influences the composition of derived products. Among the factors influencing the composition of milk, food plays an important role; it makes it possible to act in the short term and in a different way on the levels of fats and proteins (Hoden & Coulon, 1991).

The differences in composition observed between the two milks are obviously due to the species and to several factors of variation such as the mode of driving and the food. According to Haenlein and Wendorff (2006), the fat content of goat and sheep milk increases with the lactation stage while the lactose content decreases. It is commonly accepted that the milk of breeds with low dairy potential in the Mediterranean and tropical zones are generally more concentrated in fat content, total dry extract and proteins (Jenness, 1980; Voutsinas et al., 1990). According to Voutsinas et al. (1988), milk from native goat breeds has a high fat content. This content is 5.32 to 7.78% for indigenous African goats, which explains the high content of fat in the goat milk studied. Rabasco et al. (1993) explain this difference by the negative genetic correlation between the level of milk production and the fat and total nitrogen content (TNC) contents. The level of feeding corresponds to the main factor influencing the production and composition of ruminant milk (Bocquier & Caja, 2001). According to Kalantzopoulos (1993), the lowest fat contents appear at the end of the hot period (June-July). Remember that our milk samples were taken during the period between March and May.

The low pH of *Lben* is linked to the technology of its transformation and is explained by the acidifying activity of lactic acid bacteria. The acidity of milk fermented by these bacteria increases and becomes higher than the value observed in milk fresh which gives it its sour taste.

Table 1. Physico-chemical characteristics of cow's raw milk and goat's raw milk used in manufacturing of *Michouna* cheeses

	Cow's milk	**Goat's milk**
pH	6.76 ± 0.041	6.45 ± 0.02
Titratable acidity (%)	0.13 ± 0.94	0.14 ± 0.63
Dry matter (g/L)	124 ± 0.27	194.6 ± 0.17
Fat (g/L)	34 ± 0.10	52 ± 0.20
Protein (g/L)	33.4 ± 0.26	33.2 ± 0.34
Sodium chloride (g/L)	1.2 ± 0.09	1.7 ± 0.10
Ashes (g/Kg)	4.2 ± 0.04	06 ± 0.09
Calcium (mg/100mL)	1.05 ± 0.53	0.9 ± 0.35
Density	1.03 ± 0.05	1.032 ± 0.10
SNF (g/L)	89 ± 0.31	142 ± 0.43

SNF: solids-not-fat

Table 2. Physico-chemical characteristics of *Lben*

	Cow's *Lben*	**Goat's *Lben***
pH	4.55 ± 0.01	4.32 ± 0.03
Titratable acidity(%)	0.42 ± 0.32	0.47 ± 0.11
Protein (g/L)	31.3 ± 0.46	33.4 ± 0.22
Fat (g/L)	8.7 ± 0.26	24,0 ± 0.15
Ashes (g/L)	4.90± 0.03	5.2 ± 0.08
Calcium (mg/100mL)	4.00 ± 0.20	3.80 ± 0.19
Dry matter (g/L)	91.3 ± 0.28	103 ± 0.80

The pH of the goat milk noted is close to that found by Medjoudj et al. (2020) (4.39) while the pH of *Lben* prepared from cow's milk analyzed by Aissaoui Zitoune et al. (2012) is 5.05. The titratable acidities measured were on average 0.42% and 0.47% for cow and goat *Lben* respectively. We note a strong increase in the titratable acidity of *Lben* compared to that of raw milk, which indicates a significant lactic acid fermentation. These values are close to those found by Tantaoui-Elaraki et al. (1983) and Boubekri et al. (1984).

The dry matter content of *Lben* is low (91.3 g/L and 103 g/L) compared to that of raw milk, this is due to wetting during churning. The average value of the total dry extract of cow *Lben* is higher than that observed by Boubekri et al. (1984) (89 g/L), but lower than the value recorded by Medjoudj et al. (2020) for goat milk and near for the maximum value found by Tantaoui-Elaraki et al. (1983) (100.8 g/L). While the total dry extract of goat milk is higher than that of cow studied.

The fat contents of *Lben* studied are of the order of 8.7 g/L and 24 g/L for *Lben* of cows and goats respectively. The fat content of goat *Lben* is much higher than that of cow, the latter agrees with the value transferred by Boubekri et al. (1984) which is 8.93 g/L. The fat contents of *Lben* are lower than those of milk; this is explained by the extraction and separation of *Zebda* (traditional butter). According to Benkerroum et al. (1984), removing the butter granules after churning reduces the fat content by about 1.8 g/100 g. According to Benkerroum et al. (1984) and Tantaoui-Elaraki and El Marrakchi (1987), the fat content per *Lben* can range from 2 to 18 g/L.

The *Lben* protein contents analyzed are approximately 31.3 g/L and 33.4 g/L and are comparable to the values of raw milks. A slight increase in ash is noted for cow *Lben* (4.9 g/L). This increase can be explained by the minerals brought by the water during wetting, while this content was reduced in the goat's *Lben* (5.2 g/L) compared to milk; this can be explained by the losses of minerals in the *Chekoua* and their passage through the fat.

The chemical composition of *Lben* varies considerably between different localities. It is mainly due to the variation in the chemical composition of raw milk which depends on several factors (lactation period, diet and season) and the inconsistency in the manufacturing stages adopted (Tantaoui-Elaraki et al., 1984; Benkerroum & Tamime 2004; EL-Baradei et al., 2008).

3.2. Characteristics of Cheese Processing

The Table 3 presents data on the characteristics of cheese production.

The average weight relative to 1kg of the milk of three manufacturing separately shows difference (p = 2, 97939E-05). This result is reflected in the opposite direction to the weight of the whey from the milk of both breeds.

Yields of three cheeses registered are 12.3%, 14.3% and 15.6% for CA, CB and CC, respectively. Indeed, the composition of the curd and whey is directly influenced by that of milk (Rahali & Menard, 1991). The CA manufacturing non fat in dry matter is slightly higher than those of the other 2 *Michouna* which have very similar nonfat in dry matter which this cheese has lower fat levels to that of CB and CC. The ability to draining the curds (MFC) does not present significant differences between CA and CB cheeses, but CC has a higher MFC compared to the two others. Cheese *Michouna* yield made from goat's milk is higher than who's made from cow's milk. Our results were not in agreement with what is provided by Rémeuf et al. (1989), whose low cohesion goat gel can be the cause of dry matter losses in the form of "fines" in the whey which gives a lower yield in the case of cow's milk. We deduce that the yield of cheese paste is influenced by the type of raw material (milk and *Lben*), it appears that the combination of goat's milk and cow's *Lben* gives a better return than the use of raw materials the same race.

Table 3. Characteristics of cheese processing

	CA	CB	CC
Weight (g/Kg)	123	143	157
DM (%)	46.3	47.2	46.1
SNF (%)	28.8	27.7	25.6
F/DM (%)	37.5	41.1	44.4
MFC (%)	65.0	65.5	67.2
Yield (%)	12.3	14.3	15.6
Volume of whey (L)	1.19	1.18	0.98

DM: Dry Matter, F/DM: Fat/Dry Matter, SNF: Solids-Not-Fat, MFC: Moisture Free-Fat Cheese

3.3. Physico-Chemical Characteristics of Cheese

The physico-chemical characteristics of traditional *Michouna* samples are shown in Table (4).

The average pH values of studied cheeses vary between 6.04, 5.85 and 5.90 for CA, CB and CC, respectively, compared with the Hamama (1997) value for the *Jben* (4.1) are higher, but are close to those found by Abdelaziz and Ait Kaci (1992) (5.4 and 5.9) for the same cheese made from cow's milk and goat's milk respectively. Also the values recorded are higher than those reported by Leksir and Chemam (2015) (4.35-4.99) for *Klila* fresh cheese and to that communicated by Djeddar and Dahdouh (2017) (4.21 and 4.15 for cow and goat cheese respectively). Variability is related to the composition of the raw material (milk and *Lben*) and the various stages of manufacture. The averages of titratable acidity for the 3 cheeses are nevertheless lower than that of cow *Jben* (Rhiat et al., 2011). The total solids of *Michouna* of three manufacturing are 463.2g/Kg (CA), 472.2g/Kg (CB) and 461.3g/Kg (CC), these three values are lower than that proposed by Mennane et al. (2007) (456g/Kg) for *Jben* also are much higher than those recorded by Hamla and Belagroun (2019) for cow and goat *Klila* respectively, and is significantly higher than those of the values of fresh cheese, which is between 20 and 40g/100g, the amount of dry matter is mainly related to the first material and the draining time. It seems certain that the milk fat facilitates the retention of solids in cheese (Pien, 1945). The fat content of 3 cheeses ranged from 175 g/Kg, 194 g/Kg to 205g/Kg, the fat content is related to several factors namely, the raw material used, so the composition of traditional *Lben* (Table 2) is different from one preparation to another and also depends on the manufacturing conditions (diagram, expertise ...). Proteins of *Michouna* constitute the second major component with a rate of 249g/Kg, 207 g/Kg and 193 g/Kg respectively for the cheese CA, CB and CC; these levels are higher than that found by Abdelaziz and Ait Kaci (1992) to goat's *Jben*, and that provided by Hamama (1997) to cow's *Jben*. The rate of the recorded soluble nitrogen is 2.4 ± 0.63, 3.27 ± 0.56 and 3.12 ± 0.43 (g/100g). The mean levels of calcium from 3 cheeses are higher than the values advanced by Abdelaziz and Ait Kaci (1992) to cow and goat's *Jben*. The variability of calcium depends on factors, such as its amount in the raw material and the dewatering conditions. This difference mainly due to the sum of calcium provided by milk and *Lben* levels. According Brule and Lenoir (1987), about 65% of the calcium is in micellar state contributing to the formation of the curd, which explains the high levels of calcium in 3 cheeses manufacturing.

The cheeses collected have an average protein content higher than those of cheeses produced in the laboratory, the difference is not significant (t = 0.03, DOF = 25). As for fat, the values found for *Michouna* produced are higher than the average of those collected (t = 0.01, DOF = 25). The difference is not significant between the two samples with regard to ash (t = 0.02) and sodium chloride (t = 0.74, DOF = 25). The difference in values is explained by the contents of the minerals in the raw material, the amount of salt added and the method of draining. The comparison between the four samples of *Michouna* (CA, CB, CC and Ccol) shows that there is not a significant difference with regard to: pH, acidity, protein level and ash ($p > 0.05$), however, a significant difference is recorded for the level of sodium chloride, soluble nitrogen, fat and total dry extract ($p < 0.05$). This difference is due to several factors, namely: the milk of the species, the stages of manufacture, in particular the draining.

Table 4. Physico-chemical characteristics of *Michouna* cheeses

	CA (n = 5)	**CB (n = 5)**	**CC (n = 5)**	**Ccol (n = 5)**
pH	6.04 ± 0.12	5.90 ± 0.11	5.85 ± 0.02	5.85 ± 0.15
Titratable acidity (%)	0.78 ± 0.03	0.75 ± 0.05	0.81 ± 0.09	0,65 ± 0.24
Dry Matter (g/Kg)	4632 ± 0.24	472.2 ± 0.34	461.3 ± 0.12	410 ± 1.0
Fat (g/Kg)	175 ± 0.20	194 ± 0.71	205 ± 0.08	168 ± 0.89
Protein (g/Kg)	248.6 ± 0.34	207 ± 0.74	193 ± 0.52	254.8 ± 0.67
Soluble nitrogen(g/Kg)	2.40 ± 0.63	3.27 ± 0.56	3.12 ± 0.43	-
Sodium chloride(g/Kg)	2.91 ± 0.09	2.75 ± 0.01	2.62 ± 0.10	2.91 ± 0.09
Ashs (g/Kg)	4.91 ± 0.60	4.50 ± 0.03	5.13 ± 0.06	4.01 ± 0.6
Calcium (mg/Kg)	53.3 ± 0.11	55.0 ± 0.09	42.4 ± 0.21	-

CA: cheese manufacturing A; CB: cheese manufacturing B; CC: cheese manufacturing C; Ccol: cheese collected.

3.4. Microbiological Characteristics of Raw Materials

The main results of the microbiological analysis of raw materials are presented in the following Table 5.

The microbiological analysis results of traditional cheese samples are shown in Table 6. The microbiological quality of our cheese depends essentially to that of the raw materials used (Table 5).

Table 5. Microbiological characteristics of milk and *Lben*

	Cow's milk	Goat's milk	Cow's *Lben*	Goat's *Lben*
Aerobic germs (cfu/mL)	2.8×10^3	4.98×10^2	1.56×10^2	2.24×10^3
Total coliform (cfu/mL)	1.7×10^3	68.5	2.66×10^2	10
Fecal coliform (cfu/mL)	5.9×10^2	21.2	14.6	00
Staphylococci aureus (cfu/mL)	00	00	00	00
Sulphite-reducing clostridia at 46°C (cfu/mL	00	00	00	00
Salmonella (cfu/mL)	00	00	00	00
Fecal *Streptococci* (cfu/mL)	00	00	00	00
Yeasts (cfu/mL)	00	1.14×10^3	1.68×10^2	2.12×10^2
Moulds (cfu/mL)	00	00	00	00
Lactococci (cfu/mlL)	44×10^4	3×10^5	17×10^4	14×10^3
Lactobacilli (cfu/mL)	23×10^5	27×10^4	2×10^6	67×10^2
Lactic *Streptococci* (cfu/mL)	8.3×10^6	6.9×10^5	3.3×10^3	83×10^4

Table 6. Microbiological characteristics of *Michouna* cheeses

Microorganisms	CA	CB	CC	Ccol
Total aerobie Flora (cfu/g)	1.78×10^3	6.7×10^4	8.55×10^3	5.89×10^5
Total coliform (germ/g)	72.4	18.6.	9.4	3.9×10^3
Fecal coliform (germ/g)	00	00	00	19
Stahylococci (cfu/g)	00	00	00	00
Sulphite-reducing clostridia at 46°C	00	00	00	00
Salmonella (cfu/g)	00	00	00	00
Fecal*Streptococci* (germ/g)	00	00	00	23
Lactococci (cfu/g)	1.01×10^2	4×10^2	3.5×10^3	-
Lactobacilli. (cfu/g)	4×10^5	5.2×10^4	6.2×10^4	6×10^4
Lactic*Streptococci* (cfu/g)	1.78×10^3	6.7×10^4	8.55×10^3	-
Yeast (cfu/g)	00	1.14×10^3	00	3×10^2
Moulds (cfu/g)	00	00	00	00

Count of total germs milks revealed a load within standards. This flora decreases in all products, *Lben* and cheeses (Table 5 and 6). *Lben* used in CB manufacture present a load of 2.24×10^2cfu/mL and that in the whey is 1×10^2cfu/ml (Table 8), while for cheese the rate is 6.7×10^3 cfu/g. This decrease appears normal, to the lowering of the pH, the *Lben* making the unfavorable environment for large numbers of germs, and the heat treatment has an effective effect on the microorganisms which explains their decrease in the whey. In this study, the indicator flora of a non-compliance with good practice; total coliforms, bacteria attesting to fecal contamination mark their presence in milk, *Lben* and CA, the average load is respectively 1.7 x 10^3 cfu/mL; 2.66 x 10^2 cfu/mL and 1.78 x 10^3 cfu/g, for the manufacture CB, we record the presence of the flora in the milk and cheese but their absence in the whey, as to *Lben* and CC cheese their numbers are very low (10 cfu/mL and 9.4cfu/mL) which the result is considered negative. Unlike the entire fecal flora, we note the absence of fecal *Streptococci* in all samples analyzed, which may reflect the good hygienic practice during manufacturing. For *Salmonella*, *S. aureus*, and *Clostridium* sulphite-reducing were negative for all samples. The destruction of microorganisms is a function of two parameters: the temperature and duration of treatment (Alais, 1984; Vignola, 2002). In effect a heating between 55 and 85°C during a time ranging from a few minutes to an hour, depending on the methods used, the pathogens bacteria, in general the most common are killed at this temperature Yeast mark their presence in the two *Lben* at moderate levels; mostly they were recovered at the end of the fermentation step, as to mold, they are absent in the rest of the samples. Yeasts and molds with their pH optimum for growth between 4.5 and 6.5 can fully develop in the *Lben* (pH = 4.70) and there cause alterations, according to Benkerroum and Tamime (2004) different parts of *Chekoua* are the main sources of contamination because they are good niches that can easily be colonized by a variety of undesirable microorganisms. Lactic acid bacteria dominate the microflora cheese, belong to different genres and are generally recognized as non-toxic and beneficial to human health (Piard & Desmazeaud 1992). The studied cheeses samples have rates of lactic acid bacteria in the CA, CB and CC with averages of 1.78 x 10^3 cfu/ml, 6.7 x 10^4 cfu/mL and 8.55 x 10^3 cfu/mL respectively, all these results make *Michouna* of an excellent microbiological quality.

The hygienic quality of the samples collected is characterized by a density of the mesophilic flora of 5.89 x 10^5 cfu/g which remains higher than that of cheeses produced in the laboratory, but remains at the limit of standards. We record the presence of total coliforms with a load of 3.9 x 10^3 germs/g; this number is much higher than that of the samples produced, this is explained by the quality of the raw material used as well as the manufacturing conditions on farms which are frequently uncontrolled. We note the presence of coliforms and fecal streptococci in the samples collected with low levels (< 30), therefore, the result is considered negative. The research for *Salmonella*, *Staphylococci aureus* and *Clostridium* sulfite-reducers are negative for all the samples collected. The yeasts note their presence (3 x 10^2 cfu/mL) with the absence of molds. The collected *Michouna* has a microbiological quality comparable to the samples checked despite the presence of coliforms and faecal streptococci (low rate).

The microbiological quality of cheese depends on that of the starting milk, the manufacturing process it has undergone and the age of the cheese (Ercolini et al., 2009).

3.5. Physico-Chemical and Microbiological Characteristics of Whey

The physico-chemical characteristics of whey are presented in Table 7.

As for whey, their pH and acidity are close to the values communicated by Alais (1984). They contain fat levels which vary between 3 g/L and 4.3 g/L, the passage of which into the liquid phase is ensured by draining, significant rates are recorded in proteins, in particular for CB cheese. Alais (1984), specifies that the lost of water-soluble proteins in the whey have an excellent nutritional value in particular by the presence of sulfur amino acids. According to Grappin et al. (1981), all the casein does not form micelles, part of it is eliminated in the aqueous phase of milk, and this is why the percentage of casein in milk is slightly higher than the percentage of coagulable proteins. The composition of the curds and whey is directly influenced by that of milk and *Lben*. The total solids found vary between 7.07 and 10.1 g/L and are inversely proportional to the total volumes of the whey (0.985 - 1,230 L). The calcium contents are between 2.44 and 2.61 mg/L, its presence is due to its passage during draining in the liquid phase.

Table 7. Physico-chemical characteristics of whey from three manufacturers

Parameters/cheeses	whey of CA	whey of CB	whey of CC
pH	5.53 ± 0.01	4.66 ±0.07	5.54 ± 0.02
Acidity %	0.01 ± 0.00	0.029 ± 0.01	0.015 ± 0.00
Sodium chloride (g/L)	3.2 ± 0.31	2.74 ± 0.43	2.8 ± 0.81
Density	1.0 ± 0.09	1.010 ± 0.15	1.025 ± 0.05
Protein (g/L)	5.53 ± 0.23	11.9 ± 0.55	7.5 ± 0.65
fat (g/L)	3.0 ± 0.51	4.3 ± 0.53	4.01 ± 0.31
Achas (g/L)	3.54 ± 0.05	4.23 ± 0.67	3.23 ± 0.02
Calcium (mg/L)	2.61 ± 0.06	2.49 ± 0.76	2.44 ± 0.43
Total dry extract (g/Kg)	7.07 ± 0.03	9.31 ± 0.88	10.1 ± 0.95
Volum (L)	1.230 ± 0.79	1.192 ± 1.45	0.985 ± 0.78

Table 8. Microbiological characteristics of whey

	Whey A	Whey B	Whey C
Totalaerobic mesophilic bacteria (ufc/g)	1.2×10^2	1.0×10^2	2.5×102^3
Total coliform (ufc/g)	00	00	06
Fecalcoliform (ufc/g)	00	00	00
S. aureus (ufc/g)	00	00	00
Cl.S-R (ufc/g)	00	00	00
Salmonella. (ufc/g)	00	00	00
FecalStreptococci (ufc/g)	00	00	00
Yeast (ufc/g)	00	1.14×10^2	00
Molds (ufc/g)	00	00	00

The microbiological quality of whey from three manufacturing operations is presented in Table 8

Overall, the microbiological quality of whey is good. We note a low microbial load for the three samples, only the CB whey contains yeasts. However, we record the absence of the rest of the germs sought in all the samples analyzed. The hygienic quality of whey depends on the manufacturing conditions.

3.6. Sensorial Characteristics

The sensory quality of cheese depends on a number of factors related to both manufacturing technology, chemical and microbiological characteristics of the raw material used. They depend themselves many upstream factors (genetic, physiological, food ...) (Coulon, 2005).

Michouna (CA) is judged as soft cheese, with a whitish yellow color, and firm texture, elastic and spreadable. Consistency is evaluated moderately hard with a score of 3.25 ± 0.3. *Michouna* has slightly salty taste, the lactic family (subfamily lactic acid bacteria), of odors and aroma was most important in cheese (4.16 ± 1.38) followed by the animal family (subfamily cow stall- herd). For goat's *Michouna* (CB) 13 subjects state that this cheese is a soft paste, has a white color according to 15 subjects, also 10 subjects indicate that it is less firm. Goat taste registered for this cheese may be combined with some compounds of phenols which have been proposed as precursors of compounds responsible for this flavor in milk (Lopez & Lindsay, 1993). CC cheese has a firm texture and cream color, with a complex taste, due to a combination of cow's *Lben* and goat's milk aromas. A variance analysis is performed in order to study if the subject's responses are coherent and that the three cheeses were significantly different, and this for each criterion separately (Figure 2). Three cheeses have significant differences on the color plane ($p = 0,002$), texture, whose CA is firmer than the two other ($p = 0,003$), as well as the aroma with $p = 0,004$. The aromas of *Michouna* depend on those of milk and mainly on the richness of the precursor substrates of the aromatic components. Regarding the aroma of cheese, it results from equilibrium between many aromatic molecules originating of protein or lipid catalysis (Adda, 1986). The differences between the three cheeses are related to the influence of the composition of milk, the conduct of acidification, the conduct of draining. However, it is sometimes difficult to separate the effect of different factors as they are related to each other (Laithier, 2009). About the hedonic test it seems that CC cheese is the most preferred, this may be explained by the flavor resultant follows the combination of different molecules existing in both raw materials.

The cheese from goat milk is white compared to that of cow's milk with a significant difference ($p < 0.05$) and this due to the absence of β-carotene in goat milk (Chilliard, 1997), which will have an impact on goat milk products.

The gel formed by goat's milk is less firm and more brittle (Remeuf et al., 1989), the large size of goat micelles explains the weak firmness of the gel due to the negative correlation between the diameter of the micelles and firmness of the gel (Remeuf & Lenoir, 1985; Le Jaouen et al., 1990). It seems that fat also has a significant role on the texture of cheeses, in particular because of its lubricating power (Marshall, 1991). The duration of the dripping (HFD) also plays an important role in the final texture of the products, note that the subjects did not detect this difference ($p > 0.05$).

Regarding the flavor of cheeses, it results from a balance between many aromatic molecules, of lipid, protein or fermentation origin (Adda, 1986). Also terpenes, molecules specific to the plant world, have recognized odor properties in a concentrated state. They are much more abundant in certain botanical species and in particular the broadleaf weeds (Mariaca et al., 1997). These molecules pass very quickly in milk (Viallon et al., 2000) and are found in cheese, in much larger quantities when animals consume green grass or grass in natural form, rich in dicotyledons. Thus the terpene profiles of milks would make it possible to recognize the diets at the origin of these milks (Cornu et al., 2002).

The goat taste recorded for CB cheese may be due to certain conjugated phenol compounds. These compounds have been proposed as the precursors of the compounds responsible for this taste in milk (Lopez & Lindsay, 1993). However, the characteristic flavor of goat milk products is largely due to certain free fatty acids, in particular branched fatty acids (Ha & Lindsay, 1991; Le Quere et al., 1998) in particular 4-ethyl-C8 acids and 4-methyl-C8, which have a "goat" aroma (Brennand et al., 1989; Le Quere et al., 1998; Ha & Lindsay, 1991). The "goat" aroma linked mainly to fatty acidity and the lipase activity of milks (Delacroix-Buchet et al., 1996; Pierre et al., 1998). According to Morgan et al. (2001), the sensory quality of fresh cheeses was negatively correlated with the level of lipolysis of the milk used. The use of highly lipolysed milks led to the production of fresh cheeses also having high levels of lipolysis. A correlation between lipolysis and the "goat" flavor of milk has been shown (Lamberet et al., 1996; Le Quere et al., 1998; HA & Lindsay 1991).

The firmness of the cheeses studied is identical, although the CB cheese has a slightly higher dry extract and a higher Fat/Dry ratio. These results are not in agreement with the data communicated by Vassal et al. (1994) and Delacroix-Buchet et al. (1996) who indicate that the firmness of cheeses is correlated with the dry extract, Fat/Dry and pH.

Table 9. Technical card of *Michouna* cheese

Country:	**Algeria**
Cheese name:	*Michouna* cheese
Geographical demarcation:	Tébessa
Description and characteristics:	
Raw material:	Goat's or cow's raw milk
Type:	fresh soft cheese made from *Lben* and raw milk. Its color is white or whitish yellow
Composition: Fat/Dry Matter content:	Dry matter of at least: 46, 1% - 47.2% 37,5% -44,4%
Technology: Coagulation: Auxiliary coagulation: Salting: Draining: Conservation preservation: Consumption: bread	 Lactic *Lben* 12g/L (for 1L of milk) Spontaneous cold maximum 6 days spread on traditional *Galette* or

The aromas, organoleptic properties and physico-chemical characteristics of cheese depend on those of raw milk which in turn depends on the breed of animals and their type of food (Poznanski et al., 2004).

Technical Card of Traditional Michouna Cheese

Table 9 presents the first technical card of *Michouna* cheese according to physicochemical, microbiological and sensorial results.

CONCLUSION

Michouna fresh cheese usually made with goat's or cow's milk is a soft cheese, the quantities of cheese recovered change between three cheeses and vary between 123g/Kg and 157g/Kg with yields of 12.3% and 15.6%. These cheeses are characterized by a relatively acidic pH with values ranging between 5.85 and 6.04. The dry matter seems important; it varies between 461g/Kg and 472g/Kg. *Michouna* has an excellent microbiological quality. We note the absence of any pathogenic microorganisms, with dominance of lactic flora. Sensorial description shows that *Michouna* is characterized by principal aroma and odor descriptors of lactic families, and animal taste depending on the origin of the raw material used (goat or cow). From these results we can conclude that environmental conditions, type, origin and composition of milk and *Lben*, processing and sanitary conditions might have a significant influence on the microbial and physico-chemical composition of traditionally made *Michouna*. We believe that these results constitute a contribution to the identification of a regional widely consumed cheese but still unknown. Research is however needed to increase knowledge about the traditional dairy products.

REFERENCES

Abdelaziz, S. & Ait Kaci, F. (1992). *Contribution to the physico-chemical and microbiological study of a traditional Algerian cheese made from cow's milk and goat's milk: Jben*. Engineering memory in Agronomy, National Agronomic Institute, Algiers, Algeria, 67 p.

Adda, J. (1986). Flavor formation mechanisms in cheeses, in: *XXII Int Dairy Congress*, The Hague, Netherlands, 169-177.

AFNOR. (1984). *Sensory analysis, collection of French standards.* 1st edition. Paris, 159 p.

AFNOR. (1993) (a) Norme NF V04-282. (1985). *-Cheese and processed cheese: Determination of dry matter (reference method).*, p. 300-303. (b) Norme NF V04-287. (1972). Cheese: Determination of fat content, Van Gulik acid - butyrometric method. p. 323-327. (c) Norme NF V04-287. (1990). Milk: Determination of fat content. Acid-butyrometric method., p. 195- 211. In: *Collection of French standards. Quality control of food products: milks and dairy products. Physicochemical analyzes.* Afnor-dgccrf. (4ème édition.). Paris: La Défense, 562p.

Aissaoui Zitoun, O., Pediliggieri C., Benatallah, L., Lortal, S., Licitra, G., Zidoune, M. N. & Carpino, S. (2012). *Bouhezza*, a traditional Algerian raw milk cheese, made and ripened in goat skinbags. *Journal of Food Agriculture & Environment.*, *10* (2), 289–95.

Alais, C. (1984). Casein micelle and milk clotting. In *Science du lait : Principes des techniques laitières*. Paris: Ed. Sepaic, 4eme edition, 723-764.

Audigié, C. l., Figarella, J. & et Zonszain, F. (1984). *Biochemical analysis manipulation.* 1^{ére} édition, DOIN-Paris, 274.

Benkerroum, N., Tantaoui Elarki, A. & ELmarakchi, A. (1984). Hygienic quality of marrocain lben. *Microbiol Alimentation Nutrition, 2*, 199-206.

Benkerroum, N. & Tamime, A. Y. (2004). Technology transfer of some Moroccan traditional dairy products (lben, jben and smen) to small industrial scale. *Food Microbiology*, *21*, 399-413.

Bérodier, F., Lavanchy, P., Zannoni, M., Casals, J., Herrero, L. & Adamo, C. (1997). Guide to olfactory-gustatory evaluation of hard and semi-hard cheeses. *Lebensm.-Wiss. u-Technol*, *30*, 653-664.

Bocquier, F. & Caja, G. (2001). Production and composition of sheep's milk: effects of feeding. *INRA Prod. Anim.*, *14*, 129 p.

Boubekri, C., Tantaoui-Elaraki, A., Berrad, M. & Benkerroum, N. (1984). Physicochemical characterization of the Moroccan *Leben*. *Le lait*, *64*, 436-447.

Brule, G. & et Lenoir, J. (1987). Milk minerals. In milk, *the raw material of the dairy industry.* PTDCDI (Permanent Training and Development Center in the Dairy Industries), INRA, France, 520 p.

Chilliard, Y. (1997). Biochemical characteristics of goat milk lipids: comparison with cow and human milk. Nutritional value of goat milk. *French Pharmaceutical Records*, *59*, 1-51.

Cornu, A., Kondjoyan, N. & Martin, B. (2002). Towards recognition of the main diets of cows using the terpene profiles of milk. *Renc. Rech. Rum.*, *9*, 370.

Coulon, J. B. (2005). Factor of production and sensory quality of cheeses. Herbivore Research Unit, *Theix INRA*, *productions animales*, *18* (1), 49-62.

Delacroix-Buchet, A., Degas, C., Lamberet, G. & Vassal, L. (1996). Influence of AA and FF variants of goat αsl casein on cheese yield and sensory characteristics of cheeses. *Le Lait*, *76*, 217-241.

Djeddar, W. & Dahdouh, M. (2017). *Physico-chemical characteristics of traditional "Klila and Jben" cheeses prepared from cow and goat milk.* Master thesis in Biochemistry of Bioactive and Applied Molecules. University Larbi Ben Mhidi of Oum El Bouaghi, Algeria, 39 p.

Derouiche, M. & Zidoune, M. N. (2015, November). Characterization of a traditional *Michouna* cheese from the region of Tébessa, Algeria. *Livestock Research for Rural Development*, *27*(11). Retrieved from: http://www. lrrd.org/lrrd27/11/meri27229. Html.

Derouiche, M., Medjoudj, H., Aissaoui-Zitoun, W. & Zidoune, M. N. (2017). Some Traditional Cheeses Manufactures in Algeria 225-242. In Marta H F, Henriques Carlos J. D. Perira. *Cheese Production, Consumption and Health Benefits. Food Science and Technology* Nova science publishers, New York, 299.

El-Baradei, G., Delacroix-Buchet, A. & Ogier, J. C. (2008). Bacterial biodiversity of traditional *Zabady* fermented milk. *International Journal of Food Microbiology*, *121*, 295–301.

Ercolini, D., Russo, F., Ferrocino, I. & Villani, F. (2009). Molecular identification of mesophilic and psychrotrophic bacteria from raw cow's milk. *Food Microbiol.*, *26*, 228–231.

Favier, J. C. (1985). Composition of cow's milk. Mixed milk. *Cah. Nut. Diet.*, *4*, 283-291.

FAO. (1997). Manuals of Food Quality Control. Food Analysis: General Techniques, Additives, Contaminants and Food Composition n14/7. *FAO (Food and Agriculture Organization) Nutrition Paper*, Rome, 230.

Grappin, R., Jeunet, R., Pillet, R. & Le Toquin, A. (1981). Study of goat milks. I. Fat content of goat milk, nitrogenous matter and nitrogenous fractions. *Le Lait*, *61*, 117-133.

Guiraud, J. P. (1998). *Food microbiology*. Dunod, Paris, 652 p.

Ha, J. K. & Lindsay, R. C. (1993). Release of volatile branched-chain and other fatty acids from ruminant milk fats by various lipases. *Journal Dairy Science*, *76*, 677-690.

Haenlein, G. F. W. & Wendorff, W. L. (2006). Sheep milk-production and utilization of sheep milk. *In* Park, Y. W. & Haenlein, G. F. W. Edition *Handbook of milk of no bovine mammals*. Blackwell. Publishing Professional, Oxford, UK, and Ames, Iowa; USA, 137-194.

Hamama, A. (1997). Improvements of the manufacture of traditional fermented products in Morocco: case of *Jben* (Moroccan traditional fresh cheese). In: H. A. *Processing Technologies for Africa, UNIDO, Vienna, 85-102. DIRAR (éd.): Emerging Technology Series- Food.*

Hamla, H. & Belgroune, K. (2019). *Manufacture and monitoring of physicochemical and microbiological parameters of Jben and Klila made from cow and goat milk.* Master's thesis in applied microbiology. University *Larbi Ben Mhidi* of Oum El Bouaghi, Algeria, 52 p.

Hellal, A. (2001). Traditional Algerian cheeses. What future? *Review Agroligne*, *14*, 43-47.

Hoden, A. & Coulon, J. B. (1991). Control of the composition of milk: influence of nutritional factors on the quantity and levels of fat and protein. *INRA Prod Anim*, *4* (5), 361-367.

Jenness, R. J. (1980). Composition and characteristics of goat milk: a review. *J. Dairy Sci.*, *63*, 1605-1630.

Kalantzopoulos, G. (1993). State of research on goat milk in Greece. *Le lait*, *73*, 431-441.

Laithier, C., Chatelin, Y. M., Doutart, E., Barrucand, P., Duchesne, C., Morge, S., Barral, J., Cuvillier, D., Minard, L. & Leroux, V. (2009). Evaluation and control of the texture of fresh goat cheeses with lactic coagulation. *Renc. Recherche Ruminants*, *16*, 143-146.

Le Jaouen, J. C., Remeuf, F. & Lenoir, J. (1990). Recent data on goat milk and goat milk product manufacturing. XXIII *International Dairy Congress*, Octobre, 8-12, Montréal, Quebec.

Le Quere, J. L., Pierre, A., Riaublanc, A. & Demaizieres, D. (1998). Characterization of aroma compounds in the volatile fraction of soft goat cheese during ripening. *Lait*, *78*, 279–290.

Leksir, C. & chemmam, M. (2015). Contribution on the characterization of *Klila*, a traditional cheese in east of Algéria. *Livestock Research for rural development*, *27*, (5), 2015. http://www.lrrd.org/lrrd27/5/chem27083.html.

Lopez, V. & Lindsay, R. C. (1993). Metabolic conjugates as precursors for characterizing flavor compounds in ruminant's milks. *Journal Agricol Food Chemical*, *41*, 446-454.

Mariaca, R. G., Berger, T. F. H., Gauch, R., Imhof, M. I., Jeangros, B. & Bosset, J. O. (1997). Occurrence of volatile mono- and sesquiter penoids in highland and lowland plant species as possible precursors for flavor compounds in milk and dairy products, *J. Agric. Food Chem.*, *45*, 4423-4434.

Marshall, R. J. (1991). Combined instrumental and sensory measurement of the role of fat in food texture, *Food Qual. Preference.*, *2*, 117-124.

Mathieu, J. (1998). *Introduction to the physicochemistry of milk*. Lavoisier, Technical and Documentation, Paris, 220 p.

Medjoudj, H., Aouar, L., Derouiche, M., Choiset, Y., Haertlé, T., Chobert, J. M., Zidoune M. N. & Ali Adnan Hayaloglu. (2020). Physicochemical, microbiological characterization and proteolysis of Algerian traditional *Bouhezza* cheese prepared from goat's raw milk. *Analytical Letters*, *53*, (6), 905–921.

Mennane. M., Lagzouli, M., Ouhssine, M., Elyachioui, E., Berny, M., Ennouali, M. & Khedik, B. (2007). Physicochemical, microbial and sensory characterization of Morroccan *Klila. Middle East Journal of Scientific Research*, *2* (3-4), 3-4, 97.

Meyer, C. & Duteurtre, G. (1998). Milk equivalents and milk product yields: calculation methods and use. *Revue d'Élevage et de Médecine vétérinaire des Pays tropicaux*, *51*, 247-257.

Mietton, B. (1986). The preparation of cheese milks in soft cheese technology. *Industrie Alimentaire. Agriculture*, *103*, 951-963.

Morgan, F., Bodin, J. P. & Gaborit, P. (2001). Link between the level of lipolysis of goat milk and the sensory quality of cheeses made from raw or pasteurized milk. *Lait*, *81*, 743-756.

Piard, J. C. & Desmazeaud, M. J. (1992). Inhibiting factors produced by lactic acid bacteria. 2. Bacteriocins and others antibacterial substances. *Lait*, *72*, 113-142.

Pien, J. (1945). The cheese production coefficient requires a new definition. *Lait, janvier-mars*, *1*, 224-231.

Pierre, A., Le Quere, J. L., Famelart, M. H., Riaublanc, A. & Rousseau, F. (1998). Composition, yields, texture and aroma compounds of goat cheeses as related to the A and 0 variants of asl casein in milk. *Lait*, *78*, 291-301.

Poznanski, E., Cavazza, A., Cappa, F. & Cocconcelli, P. S. (2004). Alpine environment microbiota influences the bacterial development in traditional raw milk cheese. *Int. J. Food Microbiol.*, *92*, 141–151.

Rabasco, A., Serradilla, J. M., Padilla, J. A. & Serrano, A. (1993). Genetic and non-genetic sources of variation in yield and composition of milk in *Verata* goats. *Small Ruminant Res.*, *11*, 151-161.

Rahali, V. & Menard, J. L. (1991). Influence of genetic variants of β-lactoglobulin and k-casein on the composition of milk and its cheeseability. *Lait*, *71*, 275-299.

Remeuf, F. & Lenoir, J. (1985). Physicochemical characteristic of goat milk. *Revue laitière Française*, *446*, 32-40.

Rémeuf, F., Lenoir, J. & Duby. (1989). Study of the relationships between the physico-chemical characteristics of goat's milk and their ability to coagulate with rennet. *Lait*, *69*, 499-518.

Rhiat, M., Labioui, H., Driouich, A., Aouane, M., Chbab, Y., Mennane, Z. & Ouhssine, M. (2011). Comparative bacteriological study of market fresh Moroccan cheeses (Mahlabats) and cheeses produced in the laboratory. *Afrique Science*, *07*(3), 108 – 112.

Serres, L., Amariligo, S. & Petransxieni, D. (1973). *Quality control of dairy products.* Direction of Veterinary Services. The Minister of Agriculture, France.

Tantaoui-Elaraki, A., Berrada, M., El Marrakchi, A. & Berramou A. (1983). Study on the Moroccan leben. *Le Lait*, *63*, 230–245.

Tantaoui-Elaraki, A. & El Marrakchi, A. (1987). Study of the Moroccan dairy products: Lben and smen. *Mircen J.*, *3*, 211–220.

Vassal, L., Delacroix-Buchet, A. & Bouillon, J. (1994). Influence of the AA, EE and FF variants of goat asl casein on cheese yield and sensory characteristics of traditional cheeses: first observations. *Le Lait*, *74*, 89-103.

Verdier-Metz, I., Coulon, J. B., Pradel, P., Viallon, C., Albouy, H. & Berdagué, J. L. (2000). Effect of the botanical composition of hay and casein genetic variants on the chemical and sensory characteristics of ripened Saint-Nectaire type cheese. *Lait*, *80*, 361-370.

Viallon, C., Martin, B., Verdier-Metz, I.., Pradel, P., Garel, J. P., Coulon, J. B. & Berdagué, J. L. (2000). Transfer of monoterpenes and sesquiterpenes from forages into milk fat. *Lait*, *80*, 635-641.

Vignola, C. L. (2002). Milk technology science. Milk processing. Presses Internationales, *Ecole Polytechnique de Montréal*, 603 p.

Voutsinas, L. P., Delegiannis, C., Katsiari, M. C. & Pappas, C. P. (1988). Chemical composition of Boutsico ewe milk during lactation. *Milchwissenschaft*, *73*, 766.

In: Consumption and Contamination … ISBN: 978-1-53618-654-3
Editor: Egor Vagin

Chapter 2

MICROBIOLOGICAL QUALITY AND SAFETY OF DAIRY PRODUCTS IN CHEESE PRODUCTION

***Maria Schirone*[1,*], *Pierina Visciano*[1], *Alberto Maria Aldo Olivastri*[2] and *Antonello Paparella*[1]**

[1]Faculty of Bioscience and Technology for Food, Agriculture and Environment, University of Teramo, Teramo, Italy
[2]Azienda Sanitaria Unica Regionale Marche, Ascoli Piceno, Italy

ABSTRACT

In Italy, the manufacture of dairy products is an ancient tradition that allows the placing on the market of a wide variety of fresh and ripened cheeses, strongly linked to their place of origin. Their quality and safety rely primarily on the raw matter, i.e., milk that can be contaminated by microorganisms originating from udder infections, farm environment and feedstuffs, as well as milking and processing equipment. Other sources are represented by soil, faeces and bedding material attached to teats and released into milk during its collection. The microbial contamination of raw matter is a great concern for cheesemakers, as some spoilage

[*] Corresponding Author's E-mail: mschirone@unite.it.

microorganisms can affect the sensory characteristics of the products through the enzymatic alteration of milk components. However, also during the cheese production process a potential contamination can occur, due to the lack of good hygiene practices, failure of thermal treatment operations and/or temperature abuse during the storage. Moreover, spoilage microorganisms can also derive from hygiene mistakes made by the personnel working in the dairy industries. Finally, it must be highlighted that some pathogens can be present in milk and derived products, thus becoming dangerous for consumers. The Commission Regulation (EC) No 2073/2005 and its following amendments established the microbiological criteria that should be monitored by the food business operators to ensure the food safety of their products, as well as the respect of hygiene conditions during processing.

The aim of this chapter was the description of the microbiological analyses made on 128 samples of different dairy products produced by an establishment located in the Marche region, Central Italy. The results showed the compliance with the regulatory criteria, as well as the absence of pathogens, therefore demonstrating good manufacturing conditions.

Keywords: dairy products, microbiological analysis, hygiene, safety

INTRODUCTION

Cheese manufacturing is one of the most important sectors in the dairy industry, with a wide variety of fresh and ripened cheeses, very appreciated for their taste and nutritional value. Some fundamental factors of such process are coagulation (acid or by rennet), the acidification characteristics defining the mineralization level of the casein but also the moisture loss, and other technological operations, such as salting, cutting and stirring of the curds, or pressing. On the basis of the cooking temperature of the whey and curd mixture, cheeses can be classified as uncooked (< 40°C), semi-cooked (< 50°C) or cooked (> 50°C). Therefore, the whole manufacturing process is characterized by different microbiological and biochemical events, resulting in a specific type of dairy product (Almena-Aliste and Mietton, 2014). The classification of cheeses is complex due to the diversity of milk, its microbiota, the texture or composition of the final product, and some aspects linked to the production process, such as the coagulation method, the cooking temperature, or the salt addition and ripening. One of the most widely accepted approach is based on the moisture content, that allows to discriminate soft from hard cheeses.

The Commission Decision (EC) No 80/1997 classifies cheeses according to their hardness, measured by the moisture content on a fat free basis (MFFB), into 5 different subcategories (Pecorelli et al., 2020):

- soft (MFFB ≥ 68%);
- semi-soft (68% < MFFB ≤ 62%);
- semi-hard (62% < MFFB ≤ 55%);
- hard (55% < MFFB ≤ 47%);
- very hard (MFFB < 47%).

The formula for MFFB calculation is the following:

MFFB (%) = [weight of water content/ (total weight – weight of fat content)] x 100

The microbiological quality and safety of cheeses derives first of all from the raw material used in their production. The composition of the milk microbiota is very variable, with more than 100 genera and 400 microbial species, mainly Gram negative, belonging to the genera *Pseudomonas*, *Flavobacterium*, *Acinetobacter*, *Aeromonas* and various *Enterobacteriaceae*, but also Gram positive and catalase positive microorganisms, lactic acid bacteria (LAB), yeasts and molds (Montel et al., 2014; Tilocca et al., 2020). The predominant groups are streptococci and staphylococci, followed by corynebacteria and coliforms, but for technological purposes LAB are the most studied (D'Amico, 2014). Moreover, the microbiota of milk is strongly influenced by the overall management system, with direct sources of microorganisms, deriving from animals affected by systemic or udder infection, or from the skin, teats, and feces, and indirect origins associated with litter, feed, drinking water, and the staff who takes care of the animals (Verraes et al., 2015).

During the cheese manufacturing process, many differences can appear from the raw material to the final product in terms of microorganisms. The type of dairy product influences the microbial load as regards both useful and spoilage microorganisms. Psychrotrophs, coliforms, and some LAB, for instance, are responsible for the spoilage of soft, fresh cheese types, while ripened cheeses are more associated with fungi or spore-forming bacteria. Due to the high protein and lipid content, cheese alteration is mainly due to the activity of a wide range of enzymes, such as protease or lipase, causing modification of the sensory characteristics (Laslo and György, 2018). However, some intrinsic parameters may affect the growth and survival of microorganisms in cheeses, such as pH, water activity, redox potential and the presence of antimicrobial compounds produced by starter and non-starter microorganisms. LAB, used as starter cultures, contribute to the quality of fermented cheeses by improving their taste and texture and inhibiting food spoilage bacteria (Losito et al., 2014). In particular, the genera *Enterococcus*, *Lactobacillus*, *Lactococcus*, *Leuconostoc*, and *Streptococcus* can produce antibacterial substances such as H_2O_2, diacetyl and bacteriocins, as well as organic acids causing a reduction in the pH of the product (Frétin et al., 2020). Moreover, as cheeses are normally stored at a controlled temperature, the growth of pathogenic microorganisms is reduced, even if a prolonged storage can select psychrophilic or psychrotrophic spoilage bacteria, belonging to the genera *Pseudomonas*, *Psychrobacter* and *Flavobacterium* (Falardeau et al., 2019).

The aim of this chapter was the description of the analyses made in cheeses such as *Mozzarella*, *Scamorza*, *Caciotta*, *Pecorino*, etc. produced by an establishment located in the Marche region, Central Italy, and more precisely the counts of total coliforms at 37°C, *Escherichia coli*, *Enterobacteriaceae*, *Pseudomonas* spp., *Pseudomonas fluorescens*, coagulase positive staphylococci (CPS), *Listeria monocytogenes*, *Salmonella* spp. and yeasts. Most of the investigated microorganisms are reported in the Commission Regulation (EC) No 2073/2005 (Table 1) as microbiological criteria.

Table 1. Microbiological criteria for dairy products according to the Commission Regulation (EC) No 2073/2005

Food Category	Microorganism	Sampling plan		Limit
Ready to eat	*L. monocytogenes*	5	0	10^2 CFU/g
		5	0	ND*
Cheeses, butter and cream made from raw milk or milk that has undergone a lower heat treatment than pasteurization	*Salmonella* spp.	5	0	ND
Cheeses, milk powder and whey powder	Staphylococcal enterotoxins	5	0	ND
Pasteurized milk and other pasteurized liquid dairy products	*Enterobacteriaceae*	5	0	10 CFU/ml
Cheeses made from milk or whey that has undergone heat treatment	*E. coli*	5	2	10^2-10^3 CFU/g
Cheeses made from raw milk	CPS**	5	2	10^4-10^5 CFU/g
Cheeses made from milk that has undergone a lower heat treatment than pasteurization and ripened cheeses made from milk or whey that has undergone pasteurization or a stronger heat treatment	CPS	5	2	10^2-10^3 CFU/g
Unripened soft cheeses (fresh cheeses) made from milk or whey that has undergone pasteurization or a stronger heat treatment	CPS	5	2	10-10^2 CFU/g
Butter and cream made from raw milk or milk that has undergone a lower heat treatment than pasteurization	*E. coli*	5	2	10-10^2 CFU/g

Legend: * = not detected in 25 g; ** = coagulase positive staphylococci.

METHODS

A total of 128 samples of different dairy products was analyzed (Table 2). A brief description of the investigated cheeses is reported below.

Mozzarella and *Scamorza* are “pasta filata cheeses” that are manufactured primarily in the Northern Mediterranean regions, such as Italy, Greece, the Balkans and Turkey, by a thermizing and texturizing process in which the curd is dipped in hot water and worked mechanically into a plastic consistency, followed by molding into a desired shape and size (Albenzio et al., 2013). For their peculiarity, they have been traditionally grouped as a distinct cheese category (Kindstedt et al., 2010).

During the last decades, a substantial increase in *Mozzarella* production and consumption occurred worldwide, and this remarkable growth was also associated to its use as a topping for pizza (Kindstedt, 2019). It is a white rindless-pasta filata cheese, with a moisture and fat content of 45-55% and 18-20%, respectively (Cuffia et al., 2017). It is produced from cow milk through chemical acidification of the curd or by using commercial or natural whey starter cultures (Cuffia et al., 2019). In the latter condition, one of the greatest difficulties to which the primary starter cultures are subjected is the high temperature of the water during stretching. The main species of LAB that can potentially be used as probiotic cultures belong to *Lactobacillus* spp. or *Enterococcus* spp. (Reale et al., 2019).

Burrata is a fresh cheese made from cow milk and characterized by a dual structure, a bag-shaped outer part of pasta filata and an inner part of cream and pasta filata strips, usually deriving from *Mozzarella* production. It is a traditional product generally made in some regions of Central or Southern Italy, but it is known and consumed in many other countries in the world (Dambrosio et al., 2013; Rea et al., 2016).

Among fresh cheeses, the traditional Italian *Giuncata* takes its name from the common rushes used in the past to make the molds in which the cheese was drained and molded, immediately after curding at 38°C. It is obtained from cow milk and is unfermented, unsalted and cold stored for not more than 48 h (Loi et al., 2020).

Table 2. List and number (N) of the analyzed dairy product samples

Samples	Milk origin	N
Mozzarella	Bovine	46
Ricotta	Bovine or ovine	26
Cream	Bovine	19
Burrata	Bovine	14
Scamorza	Bovine	7
Caciotta	Mixture*	6
Stracciatella	Bovine	3
Giuncata	Mixture*	3
Pecorino	Ovine	2
Stracchino	Bovine	1
Mascarpone	Bovine	1

Legend: *mixture of bovine and ovine milk at different percentage.

Caciotta is one of the oldest Italian cheeses, mainly produced in the centre of country from pasteurised cow milk (sometimes with a mixture of cow and ewe milk), using thermophilic starters and dry or brine salting. It consists of two varieties, fresh or aged, with a 15 days- or 2 months- ripening period, respectively. The cured *Caciotta* has a thin and yellow rind and the dry matter is approximately 60% (Di Cagno et al., 2011).

Stracciatella is a fresh and white soft cheese produced from cow milk especially in the Central regions of Italy. In particular, raw cow milk is acidified with lactic acid and liquid rennet, and the curd is cut, stretched, and after tempering in cold water, frayed, then also fresh cream is added to the finished product (Gammariello et al., 2009).

Mascarpone is an unripened, soft cheese obtained from thermal-acidic coagulation of milk cream. The milk cream, with a minimum 20% fat content, is acidified and then coagulated by action of heat. After coagulation, the product is cooled and stored under refrigeration temperature (Carminati et al., 2001).

Stracchino is a typical Italian soft cheese obtained from cow milk heated at 37-38°C with the addition of calf rennet, culture starter and/or organic acid. After coagulation, the curd is roughly cut, left for a rest period, and then separated from the whey. To reduce pH and drain excess whey, the curd is put into plastic rectangular molds and stewed at 23 to 30°C for 5 h. The mature curd is then placed in refrigerated brine at 4°C for 4 to 5 days before aging (Manuelian et al., 2017).

Pecorino is a common name given to Italian cheeses made exclusively from pure ewe milk with a high content of fat matter. Besides different ripened *Pecorino* cheeses having in most cases a Protected Denomination of Origin, other types of *Pecorino* are characterized by a shorter aging time (20-40 days), a semi-hard consistency, and a lower flavor and aroma (Vannini et al., 2008).

Cream cheese is widely used as an ingredient in products such as flavored spreads and cheesecakes. During its production, syneresis is desired when part of the whey is removed to concentrate the curd. Fresh acid cheeses (such as cream cheese) have pH values (~ 4.6 to 5.0) that are next to the isoelectric pH of caseins (Brighenti et al., 2020).

Finally, *Ricotta* is an unripened acid-heat coagulated dairy product obtained with the use of the whey deriving from the production of different cheeses such as *Mozzarella*. More in detail, the whey is acidified with citric acid and heated at 65-70°C. Then, NaCl is supplemented and heating continues to 85-95°C. The product is hot-filled in basket molds and sealed at 75°C (Tirloni et al., 2019). The investigated microorganisms as well as the analytical methods were reported in Table 3.

Table 3. Microbiological analyses and reference method

Microorganism	**Analytical method**
Total coliforms (37°C)	ISO 4832:2006
Escherichia coli ß-glucoronidase positive	ISO 16649-2:2001
Enterobacteriaceae	ISO 21528-2:2017
Yeasts	ISO 21527-1:2008
Pseudomonas spp.	ISO/TS 11059:2009
Pseudomonas fluorescens	Pseudomonas Agar Base
Coagulase positive staphylococci	UNI EN ISO 6888-2:2004
Listeria monocytogenes	UNI EN ISO 11290-1:2017
Salmonella spp.	UNI EN ISO 6579-1:2017

RESULTS AND DISCUSSION

In the present study, the following microorganisms, i.e., *E. coli*, *P. fluorescens*, CPS, *L. monocytogenes* and *Salmonella* spp. resulted absent in all the examined samples. Low levels of *Enterobacteriaceae* and coliforms were observed in 5 out of 11 examined cheese samples, ranging from 10 to 10^2 CFU/g and 10^3 CFU/g, respectively (Tables 4 and 5). Moreover, the microbiological criteria of the Commission Regulation (EC) No 2073/2005 were always satisfactory.

Cheeses can be easily exposed to unhygienic environmental conditions not only throughout their production but also during packaging and storage, leading to cross-contamination with pathogenic or spoilage microorganisms, that can grow and survive even at refrigeration temperature.

Table 4. Number of dairy product samples positive to *Enterobacteriaceae* counts

Dairy product	10 CFU/g	10^2 CFU/g
Caciotta	1	3
Pecorino		1
Mozzarella	1	
Burrata	1	
Ricotta	1	

Table 5. Number of dairy product samples positive to total coliform counts

Dairy product	10 CFU/g	10^2 CFU/g	10^3 CFU/g
Caciotta		4	
Pecorino			1
Mozzarella	1		
Ricotta	2		
Scamorza	1	1	

Among the foodborne pathogens, *L. monocytogenes* is the most important and critical for the dairy industry, whereas *Pseudomonas* spp. as spoilage microorganism can contaminate the final product through biofilm formed on equipment and surfaces of dairy processing facilities (Carminati et al., 2019). The spoilage phenomena due to such microorganism have been shown to persist over a period of 9 months (Hyun and Lee, 2020). *Pseudomonas* spp. produce many thermo-tolerant lipolytic and proteolytic enzymes, which reduce both the quality and shelf-life of processed milk and dairy products. Moreover, *P. fluorescens* has been associated with the blue discoloration of fresh cheeses, as reported by the Italian national health authorities in *Mozzarella* imported from Germany, or in Latin-style low-acid fresh cheeses ("Queso Fresco") manufactured in the New York State from pasteurized milk (del Olmo et al., 2018). Spanu et al. (2018) described the use of a commercial biopreservative containing *Carnobacterium* spp. for the control of *Pseudomonas* spp. in *Ricotta fresca* cheese under modified atmosphere packaging. The microorganism was able to reduce *Pseudomonas* spp. growth of 1.28 log and 0.83 log after 14 and 21 days of refrigerated storage, respectively.

Carminati et al. (2019) reported the presence of *Pseudomonas* spp. in 50% of milk, 15% of cheese and 67% of processing fluid (i.e., processing water, brine and preserving liquid) of the total examined samples, with counts ranging from < 1 to 5.54 log CFU/ml for milk and from < 2 to 5.68 log CFU/g and 2.67 log CFU/ml for cheese and processing fluid, respectively. The persistency of *Pseudomonas* spp. in the dairy environment has been related to improper sanitation, biofilm formation or resistance to biocides. The ability to produce biofilm by several *Pseudomonas* spp. strains was investigated by Rossi et al. (2018) with some differences influenced by time and temperature of incubation. More in detail, these authors observed a statistically significant correlation between blue pigment production and biofilm formation, with biofilm being produced at 10°C but not at 30°C. For the production of the blue pigment by *P. fluorescens*, temperature seems more important than incubation time, probably because pigment-forming *Pseudomonas* are often adapted to low temperatures applied in production and storage (Rossi et al., 2016).

Besides *L. monocytogenes*, the major pathogenic microorganisms found in raw milk cheese are *Staphylococcus aureus*, *Salmonella* spp., *Campylobacter* spp. and verocytotoxin-producing *E. coli*. However, some production phases can affect their survival, such as salting, aging, the use of starter cultures, or the addition of ingredients able to reduce or inhibit their growth. Dupas et al. (2020) reported the addition to cheeses of aromatic herbs, spices or natural extracts for bio-preservation purposes, against oxidation for instance, but also for their antimicrobial activity. Instead, Fernández et al. (2015) described that bio-active compounds such as conjugated linoleic acid, gamma-aminobutyric acid, exopolysaccharides, oligosaccharides, and vitamins can be released in cheese by LAB, yeasts and molds. Some bio-active peptides can derive from caseins but also from whey proteins by the proteolytic activity during cheese production. An example is the serine/threonine peptide from one of the main extracellular proteins produced by *Lactiplantibacillus plantarum*, showing immunomodulatory properties when released during digestion. On the other hand, toxic compounds have been identified in dairy products due to the metabolic activity of some LAB, e.g., some biogenic amines such as tyramine, putrescine and cadaverine (Ladero et al., 2011).

Gram negative microorganisms are frequently isolated from both the surface and the cheese core, and in the latter, *Enterobacteriaceae* counts can reach 10^6-10^7 CFU/g during the first days of ripening. The presence of *Enterobacteriaceae* is very common in many traditional cheeses of the Mediterranean area (Schirone et al., 2012). They can be influenced by some physico-chemical conditions including pH, oxygen availability, water activity, proteolysis phenomena and salt content. The enzymes produced by these microorganisms can influence cheese flavor even when these bacteria are no longer viable in cheese (Chaves López et al., 2006). However, the presence of Gram negative bacteria is often used as a marker for hygiene conditions, as coliforms are indicative of fecal contamination (Coton et al., 2012).

Total coliforms were detected in different fresh and processed cheeses, ranging between $1x10^3$ and $8x10^3$ CFU/g, demonstrating the unhygienic conditions during their production process (Laslo and György, 2018).

According to Leclerc et al. (2001), coliforms can be distinguished into 3 groups, including psychrotolerant environmental coliforms, thermotolerant fecal coliforms, and ubiquitous coliforms that also include some thermotolerant coliforms. So, fecal coliforms are only a small group and *E. coli* is the only microorganism that really represents the fecal environment (Trmčić et al., 2016). As it recognizes an intestinal origin, it is considered an appropriate indicator of fecal contamination, as well as hygiene indicator. In various cheese types produced worldwide, the indicator bacteria are either not present or at levels < 10 CFU/g or < 10^2 CFU/g. Higher levels can be attributed to the use of poor-quality raw milk, or unsanitary conditions, or both. However, even if such levels increase at the initial stages of cheese-making, they tend to decline when the pH decreases, due to lactose fermentation, as well as during aging, depending on the intrinsic characteristics of the final product. Instead, the fresh cheeses, which are not aged, do not show such reduction (Metz et al., 2020).

Due to its high moisture and residual sugar content, the initial pH above 6.0 and no addition of starter cultures as competitive microflora, the fresh *Ricotta* cheese has a limited shelf-life and represents an excellent substrate for spoilage microorganisms such as *Pseudomonas* spp., *Enterobacteriaceae*, yeasts and molds (Ricciardi et al., 2019). As a matter of fact, almost all samples examined in our study were positive to these groups of bacteria, except for *Pseudomonas* spp.

Enterobacteriaceae and *E. coli* are not able to survive at the thermal treatment of the whey, therefore their presence in *Ricotta* cheese is usually attributed to secondary contamination due to low hygienic conditions during its production. Scatassa et al. (2018) reported such microorganisms in 78 out of 1295 samples of ewe ricotta cheeses in 15 years of investigation. Moreover, some fresh cheeses such as *Mozzarella* are very susceptible to spoilage processes caused by *Pseudomonas* spp., *Enterobacteriaceae* and coliforms, which have a negative impact on the texture, appearance, color, flavor and taste (Lacivita et al., 2018). Microbial counts of 10^6 CFU/g for *P. fluorescens* in *Mozzarella* and 10^5 CFU/ml for *Enterobacteriaceae* in milk have been reported as the contamination level at which the spoilage of the product starts to appear (Lucera et al., 2014). Thus, such dairy products are characterized by a short shelf-life and continuous efforts are made to extend it. Roila et al. (2019) studied the specific effect of the addition of a polyphenolic extract from olive oil into the cheese-making procedures of *Fior di latte* cheese on the inhibition of *P. fluorescens* and *Enterobacteriaceae* growth.

On the contrary, dairy products are less susceptible to fungal spoilage because they are often made with heat-treated milk, some of them are fermented products with a competitive microbiota, have an acidic pH, and naturally contain organic acids. However, visible growth of yeasts or molds on the product surface, and the production of metabolites causing off-odors and flavors, as well as visible changes in color and/or texture, have been described (Garnier et al., 2017).

In our study, yeasts were detected only in *Caciotta*, *Pecorino* and *Mozzarella* samples ranging from 10 to 10^2 CFU/g (Table 6). Several studies (Borelli et al., 2006; Lopandic et al., 2006) reported the presence of yeasts in milk and dairy products, because their intrinsic composition (i.e., protein, lipid and organic acid content), as well as other characteristics, such as lipolytic and proteolytic activities, fermentation of lactose, tolerance to high salt concentrations, low pH, low water activity and low temperatures, can favor the growth of various species of yeasts (Andrade et al., 2017). Yeasts are often present during the cheese production process and can be found in equipment, brine, and starter cultures (Andrade et al., 2019).

Table 6. Number of dairy product samples positive to yeasts counts

Dairy product	**10 CFU/g**	**10^2 CFU/g**
Caciotta	2	1
Pecorino		1
Mozzarella	1	

In conclusion, the satisfactory results obtained in the samples analyzed in this study highlighted the good hygiene practices and correct process technologies applied by the manufacturer. The consumption of dairy products has increased over the years worldwide, and the dairy industry has evolved its studies on the improvement of the shelf-life, as well as the quality and safety of cheeses. Therefore, many strategies have been implemented, such as a good hygiene management at farm level regarding both animals and milking, a time restriction between milking and subsequent processing, the maintenance of temperature as low as possible (Verraes et al., 2015) and different packaging systems, such as vacuum and modified atmosphere (Todaro et al., 2018) or with edible films and coatings (Costa et al., 2018). Moreover, the selection of suitable cleaning chemicals to remove dairy fouling (i.e., the accumulation of deposits on surfaces from process fluids) is an essential step to guarantee an efficient process (Guerrero-Navarro et al., 2019). The addition of starter cultures, generally applied in the dairy industry for technological purposes, as well as to restrict the growth of pathogens and spoilage microorganisms, is also advised for their probiotic activity and the ability to benefit the human metabolism and immunity system (Cuffia et al., 2019; Tilocca et al., 2020). As consumers' trust in safe dairy products can be damaged by a compromised manufacturing environment, it is advantageous to all food business operators to ensure the best practices and adequate control tools in the dairy industry.

REFERENCES

Albenzio, M., Santillo, A., Caroprese, M., Braghieri, A., Sevi, A. and Napolitano, F. (2013). Composition and sensory profiling of probiotic Scamorza ewe milk cheese. *Journal of Dairy Science*, 96, 2792-2800.

Almena-Aliste, M. and Mietton, B. (2014). The microbiology of traditional hard and semihard cooked mountain cheeses. *Microbiology Spectrum*, 2(1), CM-0003-2012.

Andrade, R. P., Melo, C. N., Genisheva, Z., Schwan, R. F. and Duarte, W. F. (2017). Yeasts from Canastra cheese production process: isolation and evaluation of their potential for cheese whey fermentation. *Food Research International*, 91, 72-79.

Andrade, R. P., Oliveira, D. R., Lopes, A. C. A., de Abreu, L. R. and Duarte, W. F. (2019). Survival of *Kluyveromyces lactis* and *Torulaspora delbrueckii* to simulated gastrointestinal conditions and their use as single and mixed inoculum for cheese production. *Food Research International*, 125, 108620.

Borelli, B. M., Ferreira, E. G., Lacerda, I. C. A., Franco, G. R. and Rosa, C. A. (2006). Yeast populations associated with the artisanal cheese produced in the region of Serra da Canastra, Brazil. *World Journal of Microbiology and Biotechnology*, 22, 1115-1119.

Brighenti, M., Govindasamy-Lucey, S., Jaeggi, J. J., Johnson, M. E. and Lucey, J. A. (2020). Behavior of stabilizers in acidified solutions and their effect on the textural rheological, and sensory properties of cream cheese. *Journal of Dairy Science*, 103, 2065-2076.

Carminati, D., Bonvini, B., Rossetti, L., Zago, M., Tidona, F. and Giraffa, G. (2019). Investigation on the presence of bluee pigment-producing *Pseudomonas* strains along a production line of fresh mozzarella cheese. *Food Control*, 100, 321-328.

Carminati, D., Perrone, A. and Neviani, E. (2001). Inhibition of *Clostridium sporogenes* growth in mascarpone cheese by co-inoculation with *Streptococcus thermophilus* under conditions of temperature abuse. *Food Microbiology*, 18, 571-579.

Chaves López, C., De Angelis, M., Martuscelli, M., Serio, A., Paparella, A. and Suzzi, G. (2006). Characterization of the *Enterobacteriaceae* isolated from an artisanal Italian ewe's cheese (Pecorino Abruzzese). *Journal of Applied Microbiology*, 101(2), 353-360.

Costa, M. J., Maciel, L. C., Teixeira, J. A., Vicente, A. A. and Cerqueira, M. A. (2018). Use of edible films and coatings in cheese preservation: opportunities and challenges. *Food Research International*, 107, 84-92.

Coton, M., Delbés-Paus, C., Irlinger, F., Desmasures, N., Le Fleche, A., Stahl, V., Montel, M. C. and Coton, E. (2012). Diversity and assessment of potential risk factors of Gram-negative isolates associated with French cheeses. *Food Microbiology*, 29, 88-98.

Cuffia, F., George, G., Godoy, L., Vinderola, G. and Reinheimer, J. (2019). In vivo study of the immunomodulatory capacity and the impact of probiotic strains on physicochemical and sensory characteristics: Case of pasta filata soft cheeses. *Food Research International*, 125, 108606.

Cuffia, F., George, G., Renzulli, P., Reinheimer, J., Meinardi, C. and Burns, P. (2017). Technological challenges in the production of a probiotic pasta filata soft cheese. *LWT - Food Science and Technology*, 81, 111-117.

D'Amico, D. J. (2014). Microbiological quality and safety issues in cheesemaking. *Microbiology Spectrum*, 2(1), CM-0011-2012.

Dambrosio, A., Quaglia, N. C., Saracino, M., Malcangi, M., Montagna, C., Quinto, M., Lorusso, V. and Normanno, G. (2013). Microbiological quality of Burrata cheese produced in Puglia region: Southern Italy. *Journal of Food Protection*, 76(11), 1981-1984.

Del Olmo, A., Calzada, J. and Nuñez, M. (2018). The blue discoloration of fresh cheeses: a worldwide defect associated to specific contamination by *Pseudomonas fluorescens*. *Food Control*, 86, 359-366.

Di Cagno, R., De Pasquale, I., De Angelis, M., Buchin, S., Calasso, M., Fox, P. F. and Gobbetti, M. (2011). Manufacture of Italian Caciotta-type cheeses with adjuncts and attenuated adjuncts of selected non-starter lactobacilli. *International Dairy Journal*, 21, 254-260.

Dupas, C., Métoyer, B., El Hatmi, H., Adt, I., Mahgoub, S. A. and Dumas, E. (2020). Plants: A natural solution to enhance raw milk cheese preservation? *Food Research International*, 130, 108883.

Garnier, L., Valence, F. and Mounier, J. (2017). Diversity and control of spoilage fungi in dairy products: an update. *Microorganisms*, 5(42), 1-33.

Falardeau, J., Keeney, K., Trmčić, A., Kitts, D. and Wang, S. (2019). Farm-to-fork profiling of bacterial communities associated with an artisan cheese production facility. *Food Microbiology*, 83, 48-58.

Fernández, M., Hudson, J. A., Korpela, R. and de los Reyes-Gavilán, C. G. (2015). Impact on human health of microorganisms present in fermented dairy products: an overview. *BioMed Research International*, 412714, 1-13.

Frétin, M., Chassard, C., Delbès, C., Lavigne, R., Rifa, E., Theil, S., Benoit, F. B., Laforce, P. and Callon, C. (2020). Robustness and efficacy of an inhibitory consortium against *E. coli* O26:H11 in raw milk cheeses. *Food Control*, 115, 107282.

Gammariello, D., Conte, A., Di Giulio, S., Attanasio, M. and Del Nobile, M. A. (2009). Shelf life of Stracciatella cheese under modified-atmosphere packaging. *Journal of Dairy Science*, 92(2), 483-490.

Guerrero-Navarro, A. E., Ríos-Castillo, A. G., Avila, R., Hascoët, A. S., Felipe, X. and Rodriguez Jerez, J. J. (2019). Development of a dairy fouling model to assess the efficacy of cleaning procedures using alkaline and enzymatic products. *LWT - Food Science and Technology*, 106, 44-49.

Hyun, J. E. and Lee, S. Y. (2020). Antibacterial effect and mechanisms of action of 460-470 nm light-emitting diode against *Listeria monocytogenes* and *Pseudomonas fluorescens* on the surface of packaged sliced cheese. *Food Microbiology*, 86, 103314.

Kindstedt, P. S. (2019). Symposium review: The Mozzarella/pasta filata years: a tribute to David M. Barbano. *Journal of Dairy Science*, 102, 10670-10676.

Kindstedt, P. S., Hillier, A. J. and Mayes, J. J. (2010). Technology, biochemistry and functionality of pasta filata/pizza cheese. In B. A. Law and A. Y. Tamime (Eds.), *Technology of cheesemaking* (pp. 330-359). Wiley-Blackwell, Oxford, UK.

Lacivita, V., Conte, A., Musavian, H. S., Krebs, N. H., Zambrini, V. A. and Del Nobile, M. A. (2018). Steam-ultrasound combined treatment: a promising technology to significantly control mozzarella cheese quality. *LWT - Food Science and Technology*, 93, 450-455.

Ladero, V., Rattray, F. P., Mayo, B., Martín, M. C., Fernández, M. and Alvarez, M. A. (2011). Sequencing and transcriptional analysis of the biosynthesis gene cluster of putrescine-producing *Lactococcus lactis*. *Applied and Environmental Microbiology*, 77(18), 6409-6418.

Laslo, É. and György, É. (2018). Evaluation of the microbiological quality of some dairy products. *Acta Universitatis Sapientiae, Alimentaria*, 11, 27-44.

Leclerc, H., Mossel, D. A., Edberg, S. C. and Struijk, C. B. (2001). Advances in the bacteriology of the coliform group: Their suitability as markers of microbial water safety. *Annual Review of Microbiology*, 55, 201-234.

Loi, M., Quintieri, L., De Angelis, E., Monaci, L., Logrieco, A. F., Caputo, L. and Mulè, G. (2020). Yeld improvement of the Italian fresh Giuncata cheese by laccase-induced protein crosslink. *International Dairy Journal*, 100, 104555.

Lopandic, K., Zelger, S., Bánszky, L. K., Eliskases-Lechner, F. and Prillinger, H. (2006). Identification of yeasts associated with milk products using traditional and molecular techniques. *Food Microbiology*, 23, 341-350.

Losito, F., Arienzo, A., Bottini, G., Priolisi, F. R., Mari, A. and Antonini, G. (2014). Microbiological safety and quality of Mozzarella cheese assessed by the microbiological survey method. *Journal of Dairy Science*, 97, 46-55.

Lucera, A., Mastromatteo, M., Conte, A., Zambrini, A. V., Faccia, M. and Del Nobile, M. A. (2014). Effect of active coating on microbiological and sensory properties of fresh mozzarella cheese. *Food Packaging and Shelf Life*, 1(1), 25-29.

Manuelian, C. L., Currò, S., Visentin, G., Penasa, M., Cassandro, M., Dellea, C., Bernardi, M. and De Marchi, M. (2017). Technical note : at-line prediction of mineral composition of fresh cheeses using near-infrared technologies. *Journal of Dairy Science*, 100, 6084-6089.

Metz, M., Sheehan, J. and Feng, P. C. H. (2020). Use of indicator bacteria for monitoring sanitary quality of raw milk cheeses – A literature review. *Food Microbiology*, 85, 103283.

Montel, M. C., Buchin, S., Mallet, A., Delbes-Paus, C., Vuitton, D. A., Desmasures, N. and Berthier, F. (2014). Traditional cheeses : rich and diverse microbiota with associated benefits. *International Journal of Food Microbiology*, 177, 136-154.

Pecorelli, I., Branciari, R., Roila, R., Ranucci, D., Bibi, R., van Asselt, M. and Valiani, A. (2020). Evaluation of Aflatoxin M1 enrichment factor in different cow milk cheese hardness category. *Italian Journal of Food Safety*, 9(1), 8419.

Rea, S., Marino, L., Stocchi, R., Branciari, R., Loschi, A. R., Miraglia, D. and Ranucci, D. (2016). Differences in chemical, physical and microbiological characteristics of Italian *burrata* cheeses made in artisanal and industrial plants of Apulia Region. *Italian Journal of Food Safety*, 5, 5879.

Reale, A., Di Renzo, T. and Coppola, R. (2019). Factors affecting viability of selected probiotics during cheese-making of pasta filata dairy products obtained by direct-to-vat inoculation system. *LWT - Food Science and Technology*, 116, 108476.

Ricciardi, E. F., Lacivita, V., Conte, A., Chiaravalle, E., Zambrini, A. V. and Del Nobile, M. A. (2019). X-ray irradiation as a valid technique to prolong food shelf life: the case of ricotta cheese. *International Dairy Journal*, 99, 104547.

Roila, R., Valiani, A., Ranucci, D., Ortenzi, R., Servili, M., Veneziani, G. and Branciari, R. (2019). Antimicrobial efficacy of polyphenolic extract from olive oil by product against "Fior di latte" cheese spoilage bacteria. *International Journal of Food Microbiology*, 295, 49-53.

Rossi, C., Chaves López, C., Serio, A., Goffredo, E., Cenci Goga, B. T. and Paparella, A. (2016). Influence of incubation conditions on biofilm formation by *Pseudomonas fluorescens* isolated from dairy products and dairy manufacturing plants. *Italian Journal of Food Safety,* 5(5793), 154-157.

Rossi, C., Serio, A., Chaves-López, C., Anniballi, F., Auricchio, B., Goffredo, E., Cenci-Goga, B. T., Lista, F., Fillo, S. and Paparella, A. (2018). Biofilm formation, pigment production and motility in *Pseudomonas* spp. isolated from the dairy industry. *Food Control*, 86, 241-248.

Scatassa, M. L., Mancuso, I., Sciortino, S., Macaluso, G., Palmeri, M., Arcuri, L., Todaro, M. and Cardamone, C. (2018). Retrospective study on the hygienic quality of fresh ricotta cheeses produces in Sicily, Italy. *Italian Journal of Food Safety*, 7(6911).

Schirone, M., Tofalo, R., Visciano, P., Corsetti, A. and Suzzi, G. (2012). Biogenic amines in Italian Pecorino cheese. *Frontiers in Microbiology*, 3(171).

Spanu, C., Piras, F., Mocci, A. M., Nieddu, G., De Santis, E. P. L. and Scarano, C. (2018). Use of *Carnobacterium* spp. protective culture in MAP packed Ricotta fresca cheese to control *Pseudomonas* spp. *Food Microbiology*, 74, 50-56.

Tilocca, B., Costanzo, N., Morittu, V. M., Spina, A. A., Soggiu, A., Britti, D., Roncada, P. and Piras, C. (2020). Milk microbiota: Characterization methods and role in cheese production. *Journal of Proteomics*, 210, 103534.

Tirloni, E., Stella, S., Bernardi, C., Dalgaard, P. and Rosshaug, P. S. (2019). Predicting growth of *Listeria monocytogenes* in fresh ricotta. *Food Microbiology*, 78, 123-133.

Todaro, M., Palmeri, M., Cardamone, C., Settanni, L., Mancuso, I., Mazza, F., Scatassa, M. L. and Corona, O. (2018). Impact of packaging on the microbiological, physicochemical and sensory characteristics of a "pasta filata" cheese. *Food Packaging and Shelf Life*, 17, 85-90.

Trmčić, A., Chauhan, K., Kent, D. J., Ralyea, R. D., Martin, N. H., Boor, K. J. and Wiedmann, M. (2016). Coliform detection in cheese is associated with specific cheese characteristic, but no association was found with pathogen detection. *Journal of Dairy Science*, 99(8), 6105-6120.

Vannini, L., Patrignani, F., Iucci, L., Ndagijimana, M., Vallicelli, M., Lanciotti, R. and Guerzoni, M. E. (2008). Effect of a pre-treatment of milk with high pressure homogenization on yield as well as on microbiological, lipolytic and proteolytic patterns of "Pecorino" cheese. *International Journal of Microbiology*, 128, 329-335.

Verraes, C., Vlaemynck, G., Van Weyenberg, S., De Zutter, L., Daube, G., Sindic, M., Uyttendaele, M. and Herman, L. (2015). A review of the microbiological hazards of dairy products made from raw milk. *International Dairy Journal*, 50, 32-44.

In: Consumption and Contamination … ISBN: 978-1-53618-654-3
Editor: Egor Vagin

Chapter 3

PHYSICOCHEMICAL AND MICROBIOLOGICAL CHARACTERISTICS OF ALGERIAN TRADITIONAL *LBEN* MADE FROM COW'S AND GOAT'S MILK

***Meriem Derouiche*[1,*] *and Hacen Medjouj*[2]**

[1]Laboratory of Food Engineering University of Constantine, Constantine, Algeria

[2]Institute of Sciences and Applied Techniques (I. S. T. A), University Larbi Ben Mhidi of Oum El Bouaghi, Oum EL Bouaghi, Algeria

ABSTRACT

The present study objective is to characterize one of the main fermented milk drinks consumed in Algeria "*lben*." 30 *lben* samples prepared from cow milk are collected commercially and 29 samples of *lben* prepared from goat's milk are collected from 3 farms located in Tebessa. *Lben* is characterized by its acidic taste with a pH of 4.6 and relatively low dry matter which varies between 90.6 g/L and 97.2 g/L for cow and goat respectively. The fat content is low (8.95 g/L and 13.5 g/L). The hygienic

* Corresponding Author's E-mail: meriemderouiche@yahoo.fr.

quality seems satisfactory. We record the absence of all the pathogenic micro-organisms sought and the mold. Total aerobic mesophilic flora is presented with a relatively high rate. Our results show an intense presence of lactic bacteria (*Lactococci*, *Lactobacilli* and lactic *Streptococci*).

Keywords: cow milk, goat milk, *lben*, Tebessa

1. INTRODUCTION

National consumption of milk and its derivatives is nowadays a considerable growth. They occupy an important place in the food ration of the Algerian consumer. In 2012 the recorded consumption per capita is 147 equivalents milk/year [1] which remains largely high compared to the countries of the Maghreb Tunisia and Morocco which record 83 L/year and 64 L/year [2].

Traditional Algerian dairy products, especially the fermented types, have been the pride of culinary tradition for centuries and have played a major role in the diet of communities in rural region. These products have a long history in Algeria, are traditionally made by old processes from cow's milk, goat's milk, sheep's milk or blends. They are among the first forms of milk preservation. One of these products is *lben*, which is by excellence a product of high nutritional value and taste. The Algerian *lben* is fermented milk prepared by spontaneous acidification of raw whole milk until coagulation followed by churning to recover the traditional butter. It can be manufactured throughout the year, depending on the availability of the raw material. This is the traditional dairy product most consumed in Algeria. Traditionally, churning is done in the *chekoua* (prepared from goatskin or sheep skin), with the addition of lukewarm water to facilitate the agglomeration of fat globules. The end of churning is discerned by the pieces of fat formed. Currently the traditional churning is gradually replaced by the use of electric mixers (mechanical churning) in the workshops producing *lben* for sale. *Lben* is stored for up to 3 days in cold containers in glass, plastic or clay. Indeed, the available data on physicochemical characteristics and microbiological characteristics of traditional *lben* marketed or consumed at the domestic level are, however, rare. Knowledge of these products allows the preservation of ancestral know-how and helps to support rural areas.

The objective of this study is to perform physicochemical and microbiological characterizations of Algerian *lben* made from cow's milk and goat's milk in order to identify it better.

2. Materials and Methods

2.1. Sampling

Thirty (30) samples of cow *lben* are collected at different points of sale in the town of Tebessa (East Algeria). Samples were collected as sold to the consumer and transported sterile (sterile vials) and rapidly (no more than 2 h between collection and testing) in a cooler (2 to 6°C). 29 Samples of goat *lben* samples were taken from 3 farms located in Tebessa (goat *lben* is sold only on farms). It should be noted that the *lben* (Figure 3) is prepared according to the traditional method described by Derouiche (2017) [3] churning is ensured in a *chekoua* (besides prepared from the skin of goat or sheep) (Figure 2), according to the diagram presented in Figure 1.

2.2. Physicochemical Analyses

Dry matter (DM) was determined using a forced-air, oven-drying method at 103°C for 24 h [4], the *lben* fat by Gerber method [4], total nitrogen (TN) and soluble nitrogen by the Kjeldahl method [4], for the content of protein a conversion factor of 6.38 was used. Titratable acidity was measured after titration with NaOH (N/9). The pH was determined electrometrically by means of a pH meter [5].

The Calcium was determined by flame spectrophotometry [6]. The ash content was determined by the method described by Food Agriculture Organization standard [5]. Sodium chloride assay was performed according to the method Volhart [6]. Density determination is done by the lactodensimeter.

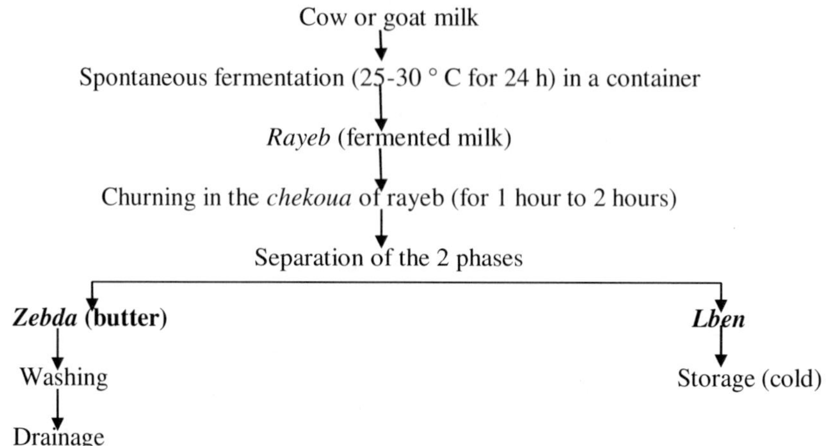

Figure 1. *Lben* manufacturing diagram [3].

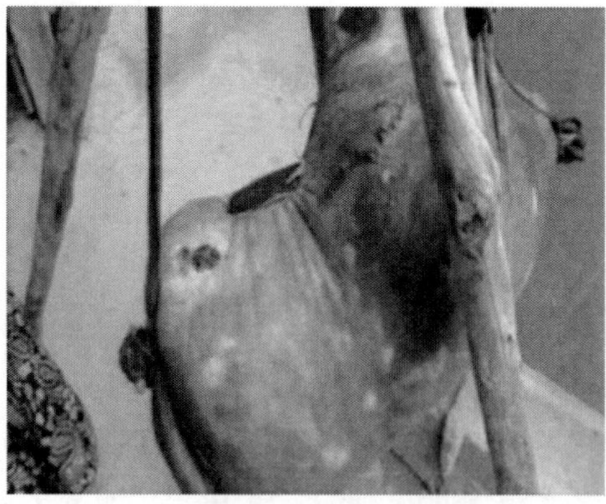

Figure 2. *Chekoua* for the preparation of *lben* [3].

Figure 3. *Lben* (prepared in *chekoua*) [3].

2.3. Microbiological Analyses

Decimal dilutions from *lben* were made using sterile quarter-strength Ringers solution as diluents. The dilutions were used for the enumeration of bacteria by conventional microbiological methods [7]. Enumeration of bacteria was carried out in triplicate. Mesophilic aerobic flora was counted on Plat Count Agar (PCA) [72 h/30°C]. *Lactococci* were enumerated on M17 agar, after incubation at 30°C for 3 days. *Lactobacilli* were enumerated on de Man Rogosa Sharpe (Difco) (MRS) agar after sowing in double layer and incubation at 30°C for 3 days, lactic *Streptococci* were counted on M17 agar after incubation at 37°C for 2 days. Yeasts and moulds were enumerated on Oxytetracycline Glucose Agar after incubation at 25°C for 5 days. *Enterobacteria* were enumerated on violet red bile glucose agar (VRBGA, Merck) for 24 h at 37°C. Total coliform are sought on BCPL incubated at 37°C for 24 to 48 h. Tubes used are those considered to be positive will be the subject of the search of fecal coliform. Fecal *Streptococci* research is carried out in two steps, first test presumption by seeding Roth incubated at 37°C for 24-48 h and the confirmatory test is done in the Litsky medium and incubated for 48 h at 37°C. The *Salmonella* requires four phases before isolation on medium SS. The *Staphylococci* were isolated after enrichment on Chapman medium and incubated at 37°C for 48 h.

Table 1. Physicochemical characteristics of *lben*

	Lben	
	Cow	Goat
pH	4.6 ± 0.03	4.61 ± 0.08
Titratable acidity (%)	0.39 ± 0.11	0.43 ± 0.41
Density	1.016 ± 0.15	1.102 ± 0.51
Dry matter (g/L)	90.6 ± 0.4	97.2 ± 0.80
Fat (g/L)	8.95 ± 0.38	13.5 ± 0.75
Protein (g/L)	30.6 ± 0.49	31.9 ± 0.22
Ash (g/L)	4.9 ± 0.02	8.5 ± 0.08
Sodium chlorides	1.71 ± 0.08	1.68 ± 0.12
Calcium (g/L)	0.7 ± 0.20	0.4 ± 0.19

2.4. Statistical Analyses

In this work, data were presented in the form of average ± standard and the comparisons were done by the Studend test processed by EXCEL.

3. Results and Discussion

3.1. Physicochemical Characteristics of *Lben*

Table 1 show the results obtained in the physicochemical analysis of the cow and goat *lben*.

The pH of two *lben* (cow and goat) are similar (4.6 and 4.6) A no significant difference is recorded between cow and goat *lben* ($t = 0.46$, $df = 57$). These values are very close to the value found by Boubekri et al. (1984) (4. 55) [8] and to that reported by El Marnissi et al. (2013) (4, 5) [9]. The low pH of *lben* is linked to the technology of its transformation and is explained by the acidifying activity of lactic bacteria whose acidity of fermented milk by these bacteria increases and becomes greater than the observed value in fresh milk which gives it its acid taste.

The titratable acidities measured were on average 0.39% and 0.34% for cow and goat *lben* respectively. These values are close and do not show a significant difference (t = 0.05, df = 57). We note a strong increase in the titratable acidity of *lben* compared to that of raw milk, which indicates a significant lactic fermentation. These values are close to those found by Tantaoui-Elaraki et al. (1983) [10] and Boubekri et al. (1984) [8].

The dry matter content of *lben* is low (90.6 g/L and 97.2 g/L) compared to that of raw milk this is due to wetting during churning. Goat *lben* is richer in dry matter. We record a significant difference between the two values (t = 7.2, df = 57). This difference can be explained by the origin and the quality of the milk (the breed, the food...), as well as by the stages of manufacture. The mean value of the dry matter of cow *lben* is higher than that observed by Boubekri et al. (1984) [8] (89 g/L) but lower than the maximum value found by Tantaoui-Elaraki et al. (1983) [10] (100.8 g/L). The fat contents of *lben* studied are of the order of 8.95 g/kg and 13.5 g/L for cow's and goat's *lben*, respectively. Its content is higher in goat *lben*, this difference is significant (t = 7.35, df = 57) and this is due to the difference of the raw material and the fat extraction stage. The fat content of goat *lben* is much higher than that of cow, the latter is consistent with the value transferred by Boubekri et al. (1984) [8] which is 8.93 g/L. The fat contents of *lben* are lower than those of the milk (33-47 g/Kg) [11], this difference is explained by the extraction and separation of the butter. Indeed according to Benkerroum et al. (1984) [12], removal of the butter granules after churning reduces the fat content by about 1.8 g/100 g. According to Benkerroum et al. (1984) [12] and Tantaoui Elaraki and El Marrakchi (1987) [13], the fat content for *lben* can range from 2 to 18 g/L.

The protein contents of *lben* are approximately 30.6 g/L and 31.9 g/L, no significant difference (t = 0.07, df = 57) and are comparable to the values of raw milk (32 and 35 g/Kg for cow milk) [11]. A slightly high ash value is noted for cow and goat *lben* (4.9 g/L and 8.5 g/L). This increase can be explained by the minerals brought by the water during wetting, compared to milk. This can contribute to the losses of the minerals their passage in the fat during churning.

The average chloride contents of our samples are 1.71 g/L and 1.68 g/L. there is not a significant difference between the recorded values ($t = 0.06$, $df = 57$). These values are close than those of whole milk (1.60 à 1.78 g/Kg) [14]. A remarkable decrease in calcium is recorded in both *lben* compared to milk, The values are lower than those reported by Desjeux (1993) (1.15 g/L) [15] and by Alais (1984) (1.30 g/L) [16] for cow's milk and goat's milk respectively, but the two *lben* do not show any significant difference ($t = 0.06$, $df = 57$).

The chemical composition of *lben* varies considerably between different localities. It is mainly due to the variation in the chemical composition of raw milk that depends on several factors (lactation period, diet and season) and inconsistency in the adopted manufacturing steps [10, 17, 18].

3.2. Microbiological Characteristics of *Lben*

The main results of the microbiological analyzes of *lben* are presented in Table 2.

The total flora is high in *lben* at 67×10^7 cfu/mL and 22×10^6 cfu/mL for cow and goat *lben* respectively. These high values of the total flora can mean the non respect of the rules of general hygiene. These rates are included in the interval found by Tantaoui-Elaraki et al. (1983) (4.1×10^7-3.8×10^{11} cfu/mL) [10]. It should be noted that the low pH of *lben* makes the medium unfavorable for a high number of germs.

Total coliform densities of two pounds are 8×10^3 and 10^2 germ/mL, while many of the fecal coliform is in the order of 7.3×10^1 and 10^1 germ/mL. Their presence even at low levels is an index of deterioration of hygienic conditions during the manufacture of *lben*. The conditions of milking, transport and preservation of the raw material and handling play a determining role in the number of coliform found. It should be noted that the presence of coliform is not necessarily an indication of fecal contamination; they are markers of general hygienic quality [18]. *Enterobacteria* are present with a rate of 4.2×10^3 cfu/mL and 3×10^3 cfu/mL this may be due to insufficient refrigeration of the raw material after milking and during its processing and even after its lbs preparation. Yeasts and molds and with their optimum pH of growth ranging between 4.5 and 6.5 [19] can perfectly develop in *lben* (pH = 4.55 and 4.32) and causing them alterations, their presence is a sign of mishandling. Yeasts mark their presence in both *lben* at levels of 39×10^2 cfu/mL for cow *lben* and 22×10^2 cfu/mL for goat, while molds are absent in all analyzed samples. The initial microbial load of the raw milk used for the manufacture of *lben*, to which a contamination of the milk is added during the manufacturing process by traditional methods probably explains the presence of the yeasts in the of cow and goat *lben*. All pathogenic microorganisms are missing. The absence of *Staphylococci* in *lben* can be explained by the strong acidification of the raw material to arrive at a lower pH, knowing that this bacterium will develop at an optimum pH of 6 to 7 [20].

Table 2. Microbiological characteristics of *lben*

Microorganisms	***Lben***	
	Cow	Goat
Total aerobie Flora (ufc/mL)	67×10^7	22×10^6
Total coliform (germ/mL)	8×10^3	10^2
Fecalcoliform (germ/mL)	7.3×10^1	10^1
Enterobacteria	4.2×10^3	3×10^2
Stahylococci (ufc/mL)	0	00
Sulphite-reducing clostridia At 46°C	0	00
Salmonella (ufc/mL)	0	00
Fecal *Streptococci* (germ/mL)	10^1	80
Yeast (ufc/mL)	39×10^2	2.2×10^2
Moulds (ufc/mL)	0	00
Lactococci (ufc/mL)	7.65×10^7	14×10^7
Lactobacilli (ufc/mL)	35.3×10^5	67×10^6
Lactic*Streptococci* (ufc/mL)	18.9×10^8	8.3×10^8

Lactic bacteria (*Lactococci*, *Lactobacilli* and lactic *Streptococci*) are present in all treated samples the latter constituting the dominant flora in all samples analyzed. Tantaoui-Elaraki et al. (1983) [10] indicate that mesophilic lactic bacteria are the main microflora responsible for lactic fermentation and aroma development in *lben* and their number in raw milk (about 10^5 cfu mL^{-1}) increased to more than 10^8 cfu mL^{-1} after 18 to 24 h. Lactic *Streptococci* and *Leuconostocs* are the main responsible agents for acidification and milk conversion in *lben* [14].

Conclusion

Lben is the most consumed dairy drink in Algeria. It is an acid taste product with a low pH (4.6). It is characterized by a low fat content due to the extraction of butter during churn (8.95 g/L and 13.5 g/L) and a relatively low dry matter content (90.6 g/L and 97.2 g/L). The total flora is high in *lben* at 67×10^7 cfu/mL and 22×10^6 cfu/mL for cow and goat *lben* respectively. Total and fecal coliforms and yeast are present at close numbers between the two *lben*. All pathogenic microorganisms are absent. Lactic bacteria are the dominant microflora in *lben* because it is responsible for the processing of milk in *lben.* Traditional fermented products have persisted over the centuries and have often evolved from a traditional and artisanal level to large-scale or industrial manufacturing. *Lben* is one of the traditional dairy products whose industrialization has seen the light in Algeria.

References

Alais, C. (1984). Casein micelle and coagulation of milk. In: *Milk Science: Principles of Dairy Techniques*, 4th ed., Lavoisier, Paris, France, pp. 723-764.

Audigie, C. L., Figarella, J. & Zonszain, F. (1984). *Manipulation D'analyse Biochimique* [*Handling Biochemical Analysis*], 1st ed.; Doin: Paris, France, 274p.

Benkerroum, N., Tantaoui Elarki, A. & El Marakchi, A. (1984). Hygienic quality of marrocain *lben*. *Microbiol. Alim. Nut.*, 2, 199-206.

Benkerroum, N. & Tamime, A. Y. (2004). Technology transfer of some Moroccan traditional dairy products (*lben*, *jben* and *smen*) to small industrial scale. *Food Microbiol.*, 21, 399-413.

Boubekri, C., Tantaoui-Elaraki, A., Berrad, M. & Benkerroum, N. (1984). Physico-chemical characterization of the Moroccan Leben. *Le Lait*, 64, 436-447.

Bouix, M. & Leveau, J. Y. (1988). The microflora responsible for the transformations. In: *Analysis and Control Techniques in food industry, Microbiological Control.* Tec. et Doc.; Lavoisier: Paris, France, Volume 3.

De Buyser, M. L. (1996). Staphylococci. In: Bourgeois, C., & Mescle, J. F. *Food Microbiology*. Eds.; *Technique et Documentation* [*Technical and Documentation*], Lavoisier: Paris, France, Volume 1, pp. 106-119.

Derouiche, M. (2017). *Milk and dairy products: diversification, frequency and modes of consumption in the Algerian tradition*. Doctoral thesis in science, Constantine University, Algeria, 189 p.

Desjeux, J. F. (1993). Goat milk and health: Nutritional value of goat milk. *Le Lait*, 73, 573-580.

El Marnissi, B., Belkhou, R., El OualiLalami, A. & Bennani, L. (2013). Microbiological and physico-chemical characterization of raw milk and its Moroccan traditional derivatives (*Lben* and *Jben*). *Les Technologies De Laboratoire* [Laboratory Technologies], 8, 100-111.

Food and Agriculture Organization (FAO). (1997). Manuals of Food Quality Control. Food Analysis: General Techniques, Additives, Contaminants and Food Composition, *FAO Nutrition Paper n 14/7*, FAO, Rome, Italy, 230p.

French Association for Standardization (AFNOR). (1993). Collections of French Standards, quality control of food products, milk and milk products, physico-chemical analyzes. Standard NF V04-287 (1990) Milk: determination of the fat content. Acido-Butyrometric method. P195-211. In: *Collection of French Standards. Quality control of food products: Milks and Dairy Products. Physicochemical analyzes. Afnor-dgccrf.* (4th Ed.); La Défense: Paris, pp. 562.

Guiraud, J. P. (1998). *Food Microbiology*. Dunod: Paris, France, 652p.

Jeantet, R., Croguennec, T., Schuck, P. & Brule, G. (2007). *Food Science*. Tec. et Doc., Lavoisier, Paris, France, 456p.

Jenness, R. J. (1980). Composition and characteristics of goat milk: A review. *J. Dairy Sci.*, 63, 1605-1630.

Kacimi El Hassani, S. (2013). Food Dependence in Algeria: Import of Powdered Milk versus Local Production, What Evolution? *Mediterr. J. Soc. Sci.*, 4, 152-158.

Larpent, J. P. (1996). Lactic acid bacteria. In: *Food Microbiology. Food Fermentation*. Tec. et Doc.; Lavoisier: Paris, France, Volume 2, pp. 3-33.

Ministry of Agriculture and Rural Development (MARD). (2013). Agricultural external trade, 2000-2012 period. In: *Agricultural Statistics Series A and B*; MADR: Algria.

Tantaoui-Elaraki, A., Berrada, M., El Marrakchi, A. & Berramou, A. (1983). Study on the Moroccan *leben*. *Le Lait*, 63, 230-245.

Tantaoui-Elaraki, A. & El Marrakchi, A. (1987). Study of the Moroccan dairy products: *Lben* and *smen*. *Mircen J.*, 3, 211-220.

In: Consumption and Contamination … ISBN: 978-1-53618-654-3
Editor: Egor Vagin

Chapter 4

CONSUMING HABITS AND ACCEPTANCE OF NEW PRODUCTS INCORPORATING DAIRY BY-PRODUCTS

Raquel P. F. Guiné * ***and Edite Teixeira de Lemos***
CERNAS-IPV, Polytechnic Institute of Viseu, Viseu, Portugal

ABSTRACT

By-products and residues from dairy industry are rich sources of nutritive components. Since their elimination brings some environmental problems, finding alternatives for their utilization assumes a particular importance, both from the economic as well as environmental points of view. In view of these challenges, the purpose of the work presented here was to examine the consumers' acceptance of bakery products incorporating whey residue, a by-product of the cheese industry. This descriptive cross-sectional study was conducted by questionnaire survey on a non-probabilistic sample composed of 299 participants aged 18 and over. Results showed that 77% of the participants eat bread on a daily basis, mostly at breakfast (65.8%) and preferred freshly baked bread. Most of them opted for bread made with rye flour (53.1%). The choice of bread was influenced by age, education and the practice of balanced diet. As for cookies, they were consumed by 56.5% of the participants on a regular

* Corresponding Author's E-mail: raquelguine@esav.ipv.pt.

basis, mainly as a snack between meals, 2 or 3 times/week. They favoured simple formulations, without fillings or flavourings. The majority of the respondents seemed to have some knowledge about whey, which was influenced by age, education, following a balanced diet and practicing exercise. Around 41% of the participants admitted they would consume bakery products made with whey residues, mostly due to their additional protein content followed by concerns about the environment and variability in taste. Considering the increasing demand for Serra da Estrela cheese, and consequent increase in by-products, their utilization in food products can be helpful to reduce water and soil pollutants while stimulating local economy.

Keywords: bread, consumers, cookies, market study, product development

1. INTRODUCTION

The domain of bakery products is very vast and diversified, including breads with many distinct characteristics as well as cakes, biscuits (crackers and cookies), among other products. All these use floor as the basic ingredient to confer structure, but the other ingredients are widely variable in nature and quantity, and even the flour used can be from very diversified sources (wheat, rice, rye, oat, corn, chestnut, acorn, carob, mushroom, etc…) (Kolawole et al., 2018; Paciulli et al., 2018; Turfani et al., 2017).

The daily consumption of bakery products is very high because these products provide basic nutrients and energy, besides being convenient to consume along the day in multiple situations including in between meals (Martins & Ferreira, 2017; Martins et al., 2017). For those reasons, bread and snacks, including cookies, are widely consumed all over the world, representing an important part of human diet, particularly as a source of energy. It has been shown that the three main meals consumed per day might not provide the nutrients needed for a healthy life, and therefore snack consumption may assume a complementary role in the nutrients and energy provision (Kolawole et al., 2018).

Nowadays, foods are regarded as much more than just vehicles for basic nutrition, and the industry has evolved towards novel foods with functional properties, intended to improve physical and mental well-being in humans as well as prevent nutrition-related diseases. In the last decades, it has been reported the rising awareness and interest of consumers for foods with improved health properties, thus expanding the marketing opportunities for these types of products, and particularly for the market of bakery products (Bimbo et al., 2017). Some studies relate the positive health effects of supplementation of bakery products, namely the ability to reduce risk of chronic diseases, antioxidant, antifungal and antimicrobial activities, anti-carcinogenic activity, among others (Acquistucci et al., 2018; Kieliszek et al., 2018; Valli et al., 2018).

Food industrial by-products and residues are rich sources of functional ingredients, like for example dietary fibre, protein, minerals, and phytochemicals, among others. Because their elimination poses some environmental challenges, to find alternative uses assumes a particular importance, both from the economic as well as environmental points of view. Their incorporation as minimally processed residues or alternatively as isolated ingredients into bakery products is in line with this strategy (Martins & Ferreira, 2017; Martins et al., 2017).

In the context of the present competitive food market, to develop new foods can be quite challenging for the food manufacturers, and they surely must guarantee that those products come to meet the consumers' expectations (Menezes et al., 2011). Besides consumers' needs and wishes, it is also important to think about the product' characteristics, technological issues and marketing strategies. Consumer surveys and market research assume a key role in guaranteeing the success of a new food product, since they provide data to evaluate the positive and less positive aspects valued by potential consumers, in time for implementing corrective actions, thus minimizing the risk of failure (Bogue et al., 2017; Derbyshire & Giovannetti, 2017).

The Serra da Estrela Cheese is a traditional Portuguese cheese with PDO (Protected Designation of Origin), natural from the Centre region of Portugal near the Serra da Estrela mountains. It is a cheese made of raw sheep milk, and the only ingredients besides milk are salt and dried thistle flower used as vegetable coagulant. Whey is the remaining liquid obtained from the making of cheese, after the pressing of the curd. This whey is used in

Portugal to make whey cheese, from which a by-product is obtained, the whey residue, which is normally discarded or used to feed cattle (Carocho et al., 2016; Reis Lima et al., 2019; Tavaria et al., 2004)

The aim of the present work was to investigate the consumer's acceptance towards the possibility of introducing in the market a new set of bakery products incorporating whey residue, thus joining the advantages linked to the functionality of some ingredients with those associated with environmental friendly waste management politics.

2. Methodology for Data Collection

To undertake this survey, the questionnaire used consisted of different parts: I. Sociodemographic data; II. Anthropometric and behavioural aspects; III. Consumption habits regarding bread; IV. Consumption habits regarding cookies; V. Acceptance of new products with whey residue. This is a descriptive cross-sectional study, based on a non-probabilistic sample of 398 participants. Data were collected between October and December 2017 in Portugal. The sample war recruited by convenience and the questionnaires were applied by direct interview, after informed consent and only to adults. All ethical issues were verified when formulating and applying the questionnaire, and confidentiality of all data collected was assured.

Body mass weight (BMI) was calculated from self-reported height and weight. Participants were classified as underweight (< 18.5 kg/m^2), normal weight (18.5-24.9 kg/m^2), overweight (25–29.9 kg/m^2) and obese (≥ 30 kg/m^2) (WHO, 2019). Sociodemographic information was also collected and age was classified into categories according to: young adults (aged between 18 and 30 years), middle aged adults (between 31 and 50 years), senior adults (between 51 and 65 years), and elderly (aged 66 years or over).

Exploratory analysis of the data included basic descriptive statistics. The relations between some categorical variables under study, were assessed with crosstabs and chi square tests with a level of significance of 5%. The strength of the significant relations found was evaluated according to Cramer's V coefficient (Witten & Witte, 2009). The influence of sociodemographic and behaviour factors on the level of knowledge was achieved by U-Mann Whitney and Kruskal-Wallis tests. The relative importance of the possible influential variables on the participants' level of knowledge about dairy by-products was assessed by tree classification analysis, following CRT (Classification and Regression Trees) algorithm with cross validation, considering minimum change in improvement of 0.0003, minimum number of cases for parent and child nodes of 20 and 10, respectively. For all data analyses it was used the SPSS-V25 from IBM Inc.

3. Results and Discussion

3.1. Sample Characterization

The sample was composed of 299 participants, from which 61.9% were women and 38.1% were men, and the average age was 33 ± 15 years. The distribution of the participants according to the ages classes defined was: young adults (18 ≤ age ≤ 30), accounting for 59.9%; middle aged adults (31 ≤ age ≤ 50), with 21.4%; senior adults (51 ≤ age ≤ 65), representing 15.7%; and finally elderly (≥ 66), being the smallest of the age groups considered with 3.0% of participants.

Most participants had completed secondary school, corresponding to 12 school years in Portugal (47.5%), while 27.1% had a university degree, and 20.1% had completed basic school (9 school years in Portugal).

Considering marital status, 57.9% of the participants were single, 35.8% were married or living together, 3.3% were separated/divorced and 3.0% were widowed. As for the profession, most of the participants (35.6%) were employed working for a third party, 34.6% were students, 11.4% worked for the government or in public companies, 6.4% were businessman, 3.4% were retired, 2.0% were unemployed, and 6.7 had other professions not specified.

Regarding the anthropometric measures presented in Table 1, in the sample at study men showed a higher incidence of overweight (52.2%) and also obesity (12.4%), while women were mostly in the class of normal weight (69.8%). Although a reporting bias may exist, since the values of weight and height were self-reported, the obtained results are in accordance with the Portuguese National Food, Nutrition and Physical activity survey 2015–2016, where weight and height were objectively measured (Oliveira et al., 2018).

When asked if the participants were satisfied with their weitgh, 50.0% of men and 52.7% of women replied affirmatively. The difference between sexes is small, even though most of the women had normal BMI, indicating that even with a normal weight, women still tend to feel unsatisfied with their body image. Frederick et al. (2016) also reported that women were more likely to be unsatisfied with body weight than men. A self-evaluation about the participants' perception regarding the frequency of practicing a healthy diet showed that 7.0% of men never did it, 20.2% said rarely, 36.6% said sometimes, 29.8% many times and 6.1% replied always. More women believed to practice frequently a healthy diet than men (3.3% never, 8.2% rarely, 50.0% sometimes, 31.0% many times and 7.6% always). In the study by Bissonnette-Maheux et al. (2018) was found that women rely frequently on healthy eating blogs made available by nutrition and dietetics specialist and this might explain why they tend to feel more informed and more able to make healthier food choices.

Table 1. Anthropometric measures according to sex

Characteristic	Women	Men	Global
Height (m)	1.63 ± 0.06	1.76 ± 0.07	1.67 ± 0.09
Weight (kg)	61.44 ± 10.82	80.54 ± 11.76	68.75 ± 14.54
BMI[1] (kg/m^2)	23.09 ± 3.87	26.13 ± 3.52	24.26 ± 4.02
Underweight (BMI < 17.9)	5.5%	0.9%	3.7%
Normal weight (18.0 ≤ BMI ≤ 24.9)	69.8%	34.5%	56.3%
Overweight (25.0 ≤ BMI ≤ 29.9)	19.8%	52.2%	32.2%
Obese (BMI ≥ 30.0)	4.9%	12.4%	7.8%
Total	100.0%	100.0%	100.0%

[1]BMI = Body Mass Index

Most participants, 46.6%, practiced an insufficient level of structured physical activity, corresponding to only once a week, and 24.5% assumed they did not practice exercise at all. Only 24.5% practiced exercise 2 or 3 times/week and a minority, 4.4%, surpassed 3 times/week. The young adults follow a pattern very similar to that described for the whole sample (never: 26.8%, 1 time: 43.0%, 2-3 times: 24.6%, +3 times: 5.6%), and the adults also did not differ much, with the majority practicing once/week (51.6%). It is interesting that senior adults seemed more active, with the percentage of participants that never practiced exercise reduced to 17.0%, while for the elderly that percentage considerably increased, 44.4% never practiced exercise. According to Kendrick et al. (2018) older people tend to achieve insufficient levels of physical activity and Burnet et al. (2019) suggested that incentives to a greater involvement in the engagement of physical activity should include social, cognitive, personality, environmental or socioeconomic factors (Rhodes et al., 2009).

3.2. Consuming Habits Regarding Bread

Eating bread on a regular basis was reported by 81.6% of participants, and most of them (77%) consumed bread every day, with the rest (21.4%) consuming bread once or twice a week. A minority, 1,6%, consumed bread only occasionally. This supports earlier findings of Gül et al. (2003), who reported similar purchase and consumption rates of bread in Adana Province, Turkey.

Most participants reported consuming bread for breakfast (65.8%) and snacks, while 28.4% consumed it with lunch and dinner. This is in accordance with Aljobair (2017), where 49.42% consumed bread at breakfast. Women consumed bread preferably at breakfast (70.8%), while men did it throughout the day (breakfast 57.3%, other meals and snacks 42.7%). Bread consumption habits are shown in detail in Table 2.

Table 2. Frequency and moments of bread consumption, according to gender

Bread consumption		Global		Women		Men	
		Yes (%)	No (%)	Yes (%)	No (%)	Yes (%)	No (%)
Frequency	Daily	77.0	23.0	76.6	23.4	77.8	22.2
	Two or three times/week	18.9	81.1	18.2	81.8	20.0	80.0
	Once/week	2.5	97.5	3.9	96.1	0.0	100.0
	Sporadically	1.6	98.4	1.3	98.7	2.2	97.8
Moments[1]	Breakfast	65.8	34.2	70.8	29.2	57.3	42.7
	Morning/afternoon tea	52.3	47.7	55.2	44.8	47.2	52.8
	With meals	28.4	71.6	19.5	80.5	43.8	56.2

[1]Multiple options were possible

Table 3 presents the preferences regarding the preparation and constitution of the bread the participants consume. 70% consumed freshly made bread, while 15.2% favoured homemade bread and only 3.3% opted for prepacked bread. Men were more interested in daily made bread than women (77.8% vs 66.7%).

Table 3. Preferences regarding bread type and composition, according to gender

Bread preferences		Global		Women		Men	
		Yes (%)	No (%)	Yes (%)	No (%)	Yes (%)	No (%)
Type	Fresh (daily made)	70.8	29.2	66.7	33.3	77.8	22.2
	Prepacked	3.3	96.7	3.9	96.1	2.2	97.8
	Home made	15.2	84.8	16.3	83.7	13.3	86.7
	No preference	10.7	89.3	13.1	86.9	6.7	93.3
Composition[1]	Wheat flour	46.1	53.9	40.0	60.0	56.7	43.3
	Rye flour	53.1	46.9	55.5	44.5	48.9	51.1
	Corn flour	26.5	73.5	34.2	65.8	13.3	86.7
	Mixture flour	24.1	75.9	27.1	72.9	18.9	81.1
	Whole cereal	32.4	67.6	31.2	68.8	34.4	65.6
	With seeds flour	28.6	71.4	33.5	66.5	20.0	80.0
	Reduced salt	10.6	89.4	12.3	87.7	7.8	92.2

[1]Multiple options were possible

Table 3 also shows that women preferred homemade bread (16.3% vs 13.3%). Most participants preferred bread made with rye flour (53.1%), followed by wheat flour (46.1%). Healthier options were less popular. Breads made with whole grain flour were selected by 32.4% of participants and breads with seeds were consumed by 28.6%, while only 11,82% opted for breads with lower salt content. When compared to their male counterparts, female respondents seemed to pay more attention to the composition of bread, preferring those with healthier flours. In a study conducted in Finland, rye bread consumption was found to be consistently associated with rural places of residence and lower educational attainment. In contrast, white bread was more widely consumed in urban areas, by people with lower education levels (Prättälä et al., 2001). In another study, conducted in a young population (14-26 years of age), Šereš et al. (2017) reported that white bread was consumed by all respondents, elementary school students in particular, as a result of family impact and tradition. As their education level rises, the young population becomes more aware of health related benefits provided by whole grain products. Female respondents appeared to be more interested in healthier dietary practices, in comparison with men (Šereš, Simović, et al., 2017).

Statistical analysis of the factors which could influence frequency of consumption and type of bread preferences is summarized in Table 4. The only statistically significant difference found in the frequency of bread consumption was between sedentary participants and those who exercise ($p < 0.05$). These results differ from those found in a national survey (sample of 5819 Portuguese adults) where the amount consumption of bread seemed to be associated with age (Lopes et al., 2017). An earlier cross-sectional study made with a representative sample of Portuguese adults by Moreira and Padrão (2004) demonstrated that the consumption of bread (in both genders) decreased with higher levels of education. They hypothesise that this could be because more educated individuals tend to avoid foods such as bread, that are considered fattening or rich in energy. The main factors affecting the preferred type of bread are age, education and practicing a balanced diet ($p < 0.05$). These results are in accordance with those found by Sandvik et al. (2014) in a Swedish population, and by Šeres et al. (2017)

in a young Serbian population, who also found a correlation between the type of bread consumed and age, education and healthy-lifestyle-related factors.

Table 4. Factors influencing the habits and preferences of bread consumption

Demographic and behavioural factors	Frequency of consumption[1]		Preference for type of bread[1]	
	p-value	Cramer´s V	p-value	Cramer´s V
Age group	0.257	0.124	0.004	0.181
Gender	0.271	0.127	0.262	0.128
Education	0.185	0.164	0.031	0.191
Marital status	0.422	0.112	0.198	0.130
Employment	0.156	0.194	0.146	0.195
Exercise	0.024	0.162	0.900	0.076
BMI class	0.179	0.132	0.619	0.100
Balanced diet	0.055	0.168	0.002	0.208
Weight satisfaction	0.514	0.097	0.528	0.091

[1]Chi-square test, with a level of significance of 5%

3.3. Consuming Habits Regarding Cookies

Cookies were consumed on a regular basis by 56.5% of the participants, a lower percentage when compared to the consumption of bread. Despite the fact that in the past few years many types of biscuits or cookies have been designed for multiple occasions and for various type of consumers, the participants still consume them mainly between meals and 2 or 3 times/week (Table 5). These types of baked goods are typically high in fats and sugars and for this reason health professionals recommend that they should only be eaten in small quantities as part of a varied diet (Astrup et al., 2006). In developing countries, however, these types of baked products are typically used as a positive food vector when implementing food fortification programmes. As discussed by Mannar and Sankar (2004), cookies are easy to distribute, easy to prepare and fortify and have a long shelf life for use in long-term feeding programmes.

Table 5. Frequency and moments of consumption of cookies, according to gender

Cookies' consumption		Global		Women		Men	
		Yes (%)	No (%)	Yes (%)	No (%)	Yes (%)	No (%)
Frequency	Daily	30.2	69.8	33.9	66.1	21.6	78.4
	Two or three times/week	46.7	53.3	45.8	54.2	49.0	51.0
	Once/week	14.2	85.8	12.7	87.3	17.6	82.4
	Sporadically	8.9	91.9	7.6	92.4	11.8	88.2
Moments[1]	Breakfast	13.6	86.4	11.9	88.1	17.6	82.4
	Morning/ afternoon tea	70.4	29.6	76.3	23.7	56.9	43.1
	With meals	36.7	63.3	34.7	65.3	41.2	58.8

[1]Multiple options were possible

As shown in Table 6, which summarizes the preferences of the participants regarding cookies, both men and women tend to prefer simple formulations, without fillings or flavourings. As a second option, however, women prefer flavoured cookies, while men favoured cookies filled with cream. Cookies without sugar are preferred only by 11.2% of participants, with women favouring this option more than men (13.6% vs 5.9% respectively). Trivedi (2016), evaluating the consumer behaviour perspective on nutritious biscuits in India, found that preference was given primarily to cookies with cream followed by glucose type of biscuits. In baked goods such as cookies and cakes, sugar is one of the main components, contributing up to 30-40% of the total recipe. Due to increasing health concerns associated with the wide availability of energy-dense foods and excessive caloric intake, extensive research is ongoing in order to replace sugars with healthier alternatives. However, the present results show that these reformulations tend to be less favoured by consumer, as only a minority of respondents seemed to prefer sugar-free cookies.

This could be explained by the fact that the reduction of sucrose in biscuits or cookies can cause detectable losses in appearance, rheological and textural properties of the products, flavour, and mouthfeel of the final foods (Mariotti & Lucisano, 2014).

Table 6. Preferences regarding cookies, according to gender

Cookies' preferences		Global		Women		Men	
		Yes (%)	No (%)	Yes (%)	No (%)	Yes (%)	No (%)
Type[1]	Basic	56.8	43.2	58.5	41.5	52.9	47.1
	With flavours	48.5	51.5	52.5	47.5	39.2	60.8
	Filled with fruit	18.9	81.1	20.3	79.9	15.7	84.3
	Filled with cream	42.0	58.0	38.1	61.9	51.0	49.0
	Home made	20.7	79.3	22.9	77.1	15.7	84.3
	Sugar free	11.2	88.8	13.6	86.4	5.9	94.1
	Others	5.9	94.1	6.8	93.2	3.9	96.1

[1]Multiple options were possible

In the sample at study, no statistically significant differences in cookies' consumption were found when looking at the demographic and behavioural characteristics of the respondents (Table 7). These results conflict with those of Aranceta et al. (2003), who found that children and young people from a lower socioeconomic background and those whose mothers had a lower level of education showed a higher consumption of biscuits and cookies.

3.4. Knowledge about Valorization of Dairy By-Products

Nowadays, it is increasingly important to undertake environmental-friendly approaches in production industries. Dairy industries are no exception, and its by-products are frequently rich in components with nutritional value, such as proteins, minerals, casein, among others. To assess the respondents' knowledge about the valorisation of dairy by-products, a number of statements were included in the questionnaire. The respondents were asked to rate their level of agreement as presented in Table 8.

Table 7. Factors influencing the consumption of cookies

Demographic and behavioural factors	Frequency of consumption[1]	
	p-value	Cramer's V
Age group	0.405	0.136
Gender	0.366	0.137
Education	0.881	0.133
Marital status	0.523	0.126
Employment	0.866	0.167
Exercise	0.373	0.139
BMI class	0.126	0.167
Balanced diet	0.232	0.173
Weight satisfaction	0.634	0.101

[1]Chi-square test, with a level of significance of 5%

Table 8. Frequency of responses obtained for the statements used to evaluate the knowledge about valorisation of dairy by-products

Statement	Scale 1 (totally disagree) to 5 (totally agree)					No opinion
	1	2	3	4	5	0
Residues from the food industry can be used for human consumption under certain conditions	7.0%	7.0%	21.1%	18.7%	29.8%	16.4%
Is possible to enrich foods with bioactive compounds from food industry waste (e.g., fiber, protein, antioxidants, vitamins, etc.)	2.3%	3.0%	22.7%	21.1%	36.5%	14.4%
Whey is a by-product of cheese production	3.7%	3.7%	7.7%	18.4%	47.5%	19.1%
Whey is mainly constituted by water	3.0%	5.0%	15.4%	22.1%	36.6%	17.8%
Whey contains protein	1.0%	3.0%	12.0%	20.1%	40.5%	23.4%
Whey contains minerals	2.0%	2.3%	17.7%	16.7%	36.8%	24.4%
Whey contains fat	5.4%	7.0%	17.4%	16.4%	32.1%	21.7%
The recovery of waste and by-products is economically beneficial	1.3%	2.0%	13.4%	23.1%	41.8%	18.4%
The recovery of waste and by-products brings environmental benefits	2.0%	1.7%	13.0%	21.1%	43.5%	18.7%
Whey is a pollutant if discarded into the environment	16.1%	7.7%	12.4%	11.7%	22.4%	29.8%

The majority of the participants expressed agreement with the items about the origin of whey and its composition (Table 8). These findings seem to contrast with those found in the 2014 Consumer Whey Protein Tracker Study, a research project backed by the U.S. Dairy Export Council (USDEC) and the Whey Protein Research Consortium were only about 30 percent of the population identified milk as the source of whey protein (Lagrange, 2015). Table 8 also shows that respondents are generally in agreement with questions concerning the environmental problems. Moreover, and consistently with the health concerns demonstrated in other parts of the survey, more than 50% strongly agree with the possibility of enriching foods with bioactive compounds. To our knowledge, no other studies were conducted with a focus on these questions. Nevertheless, the consumer's expectations for products that are simultaneously palatable and healthy is challenging for the development of novel foods and/or functional food products. A recent review of Iriondo-DeHond et al. (2018) provides an overview of the current trends in the use of food by-products, and also shows a gap in sensory and consumer expectations.

Based on the responses of the participants, a new variable was created to account for the level of knowledge, including all items and calculated as their sum. This variable varied between a minimum value of 10 and a maximum of 50, and was categorized into classes as follows: very low knowledge (1-10 points), low knowledge (11-20 points), medium knowledge (21-30 points), good knowledge (31-40 points), very good knowledge (41-50 points). Table 9 reveals that most respondents showed a positive level of knowledge, with 35.9% of the participants having a good knowledge and 26.2% a very good knowledge about the dairy by-products.

Table 9. Frequencies obtained for the level of knowledge

Level of knowledge	Points	Frequency (%)
Very low	1 – 10	4.1
Low	11 – 20	11.0
Medium	21 – 30	22.8
Good	31 – 40	35.9
Very good	41 – 50	26.2

The obtained results showed that having a balanced diet influenced the knowledge about whey, as shown in Table 10 ($p < 0{,}05$). These results suggest that healthy habits and motivations are consistent with a better knowledge of ingredients. Several studies demonstrated that people with a higher awareness of ingredients and functional foods in general tend to be more willing to buy these foods (Bornkessel et al., 2014; Lu, 2015).

Table 10. Factors influencing knowledge about whey

Demographical and behavioural factors	**Level of knowledge**	
	p-value	**Type of test**[1]
Age group	0.093	Kruskal Wallis
Gender	0.929	U Mann Whitney
Education	0.586	Kruskal Wallis
Marital status	0.040	Kruskal Wallis
Employment	0.809	Kruskal Wallis
Exercise	0.091	Kruskal Wallis
BMI class	0.085	Kruskal Wallis
Balanced diet	0.016	Kruskal Wallis
Weight satisfaction	0.142	U Mann Whitney

[1]Level of significance considered: 5%

Marital status also influenced knowledge about whey ($p < 0.05$). This is in accordance with a study by Parmenter et al. (2000), who found that married or living as married individuals achieved slightly higher nutritional knowledge than those who were single, separated, divorced or widowed. Later studies also supported these findings (Hakli et al., 2016; Mirmiran et al., 2010). Marital status seems to impact the lifestyle of an individual as a whole anywhere in the world (Wilson, 2001). The decision tree procedure creates a tree-based classification model, classifying cases into groups or predicting values of a dependent variable (output) based on the values of the independent variables (inputs). The procedure may include validation tools for exploratory or confirmatory classification analysis. In this work, tree classification was obtained with CRT Algorithm to identify discriminant variables and their interactions on the dependent variable "level of knowledge". Two separate sets of dependent variables were used, the first regarding the sociodemographic factors (Age group, Gender, Education,

Civil State) and the second considering the behavioural aspects (Balanced diet, Exercise, BMI class, Weight satisfaction). Figure 1 shows the classification tree that relates sociodemographic variables with level of knowledge. This tree contains 9 nodes (5 are terminal) and 3 levels. From the variables considered, only three were included in the analysis, meaning that the variable gender was not found sufficiently discriminant. Results indicate that age appears as the first discriminant variable, followed by education, this last being discriminant for all age groups, although with a different influence. The risk estimated for the solution with cross validation using CRT algorithm was 0.683, with a standard deviation of 0.027. The targeted level of knowledge was very high, with an incidence of 26.2% in the initial node, but increasing to 33.7% in node 1, i.e., for adults between 31 and 65 years old, and being highest, 44.8%, for the subgroup of participants who had completed a university level of education. Figure 2 shows the classification tree obtained for the relation between behavioural variables (Balanced diet, Exercise, BMI class, Weight satisfaction) and level of knowledge. The number of nodes in this tree is higher, 13, from which 7 are terminal, with a depth of 5 levels. In this case all the variables considered were included in the analysis, and therefore all of them were considered to have a discriminant capacity over the level of knowledge. The risk estimated for the solution, with cross validation using CRT algorithm, was 0.693 with a standard deviation of 0.027. To practice a balanced diet appeared as the first discriminant variable, but for the participants who never or rarely followed a balanced diet no more influential variables were found, being this a terminal node and the majority (29.4%) revealed a low level of knowledge. For the group of people who tend to follow more regularly a balanced diet, 38.1%, revealed a high level of knowledge and the next discriminant factor was exercise. For the participants who tend to practice low levels of physical activity a further discriminant variable is again the dietary pattern, and BMI appears as discriminant for those who practice a balanced diet at least sometimes, but who revealed a high level of knowledge (38.1%, node 5). In the next and last depth level, while for the participants with low and normal weight the next discriminant was exercise, for those with overweight or obese the discrimination came from satisfaction with bodyweight.

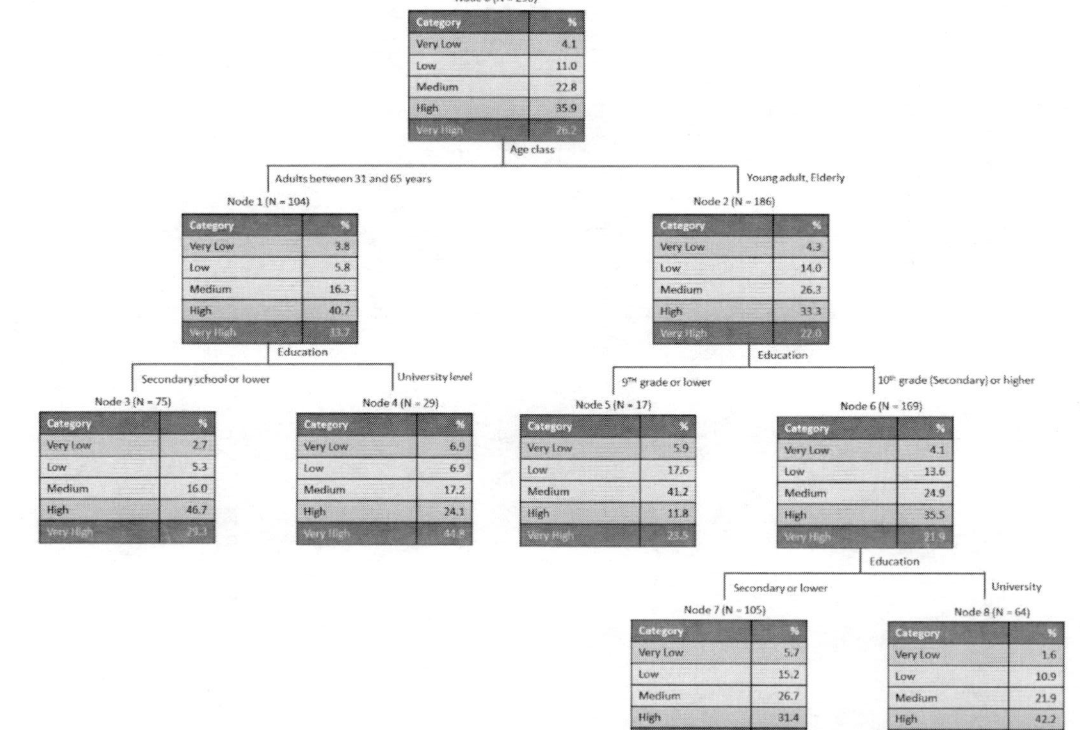

Figure 1. Classification tree for the relation between sociodemographic aspects and the level of knowledge.

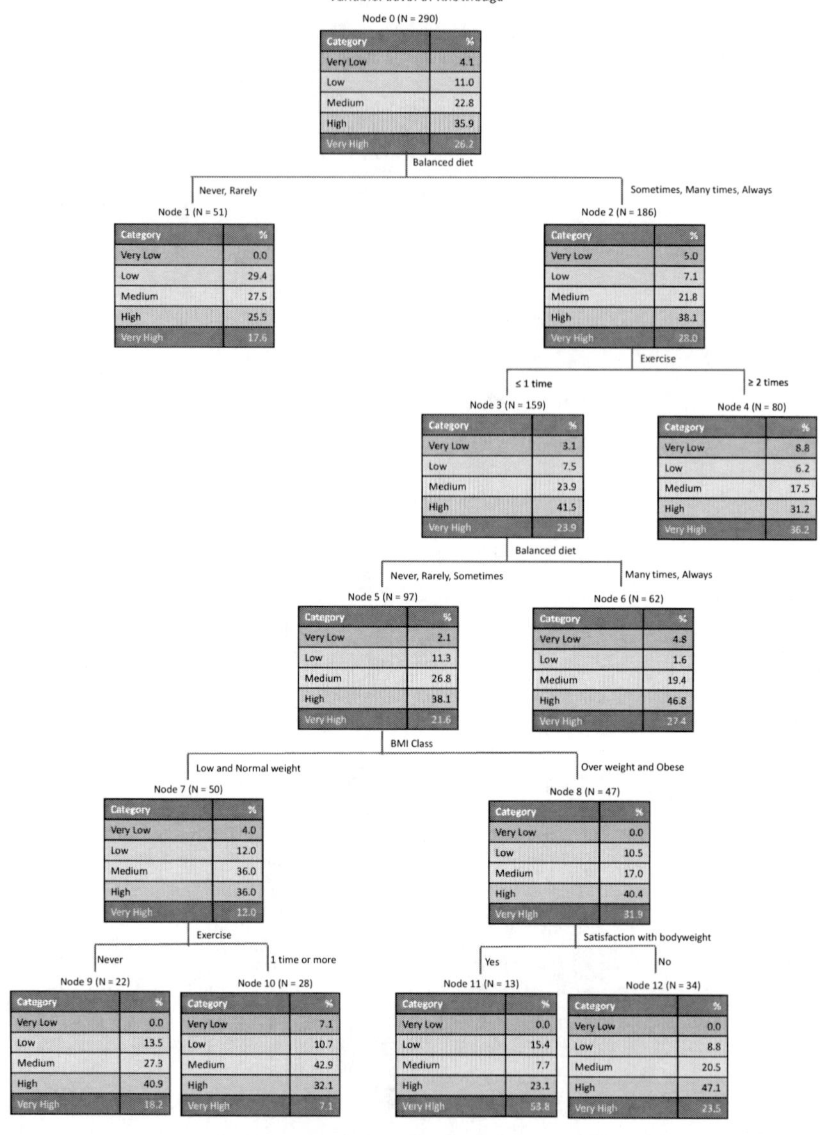

Figure 2. Classification tree for the relation between behavioural aspects and the level of knowledge.

3.5. Acceptance of New Products with Whey Residue

When the participants were asked if they would consume bakery or pastry products incorporating whey residue, 40.7% answered yes and 59.3% admitted they would not. When asked if they would consume baked goods incorporating whey residue, 40.7% answered yes and 59.3% answered no. Among the reasons for consuming this type of product, the most relevant factors were their high protein content (valued by 62.6% of the participants), followed by environmental concerns (29.8%) and variability in taste (28.6%) (Figure 3). Recent research on consumer perception and behaviours towards whey protein products has been limited to whey protein-added snack, yogurt, and meal replacement products (Childs et al., 2008; Lee et al., 2006). The interest in the protein content can be explained by the health concerns expressed by the participants. The work of Childs et al. (2008) seems to support this hypothesis, as most of the subjects in their study identified foods high in protein as healthier.

The present results also revealed that consumers have growing environmental concerns, namely with industrial practices which increase soil and water pollution, the depletion of natural resources, and the destruction of delicate ecosystems and biodiversity, mirroring the findings of a study conducted by National Geographic and GlobeScan in 2014 in 18 countries. They found that more than 50% of consumers are very concerned about environmental problems and about 40% believe that environmental problems are having a negative impact on their health. They also associated environmental concerns with a rise in the consumption of more sustainable food products (Malmqvist & Wham, 2016).

Additionally, the present results bring up the importance of taste in the consumption of baked goods incorporating whey residue. Taste influences the food decision process. If taste is not good, the product won't be consumed. Because the effects of functional foods are not yet measurable, the functionality plays a minor role than taste, thus we might consider that, regardless of innovative tendency, consumers adopt the product proposed to them for hedonic reasons. Others have found similar results (Barrena-Figueroa & Garcia-Lopez-de-Meneses, 2012).

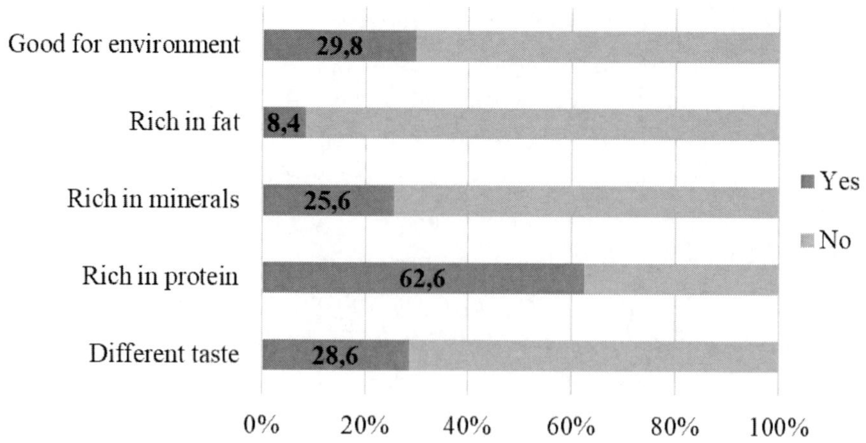

Figure 3. Reasons for acceptance of the new products.

Table 11. Factors influencing the acceptance new products with whey residue

Personal and behavioural factors	Acceptance[1]	
	p-value	Cramer´s V
Age group	0.172	0.130
Gender	0.535	0.036
Education	0.065	0.187
Civil State	0.478	0.091
Employment	0.109	0.199
Exercise	0.108	0.143
BMI class	0.816	0.057
Balanced diet	0.105	0.161
Weight satisfaction	0.019	0.137

[1]Chi-square test, with a level of significance of 5%

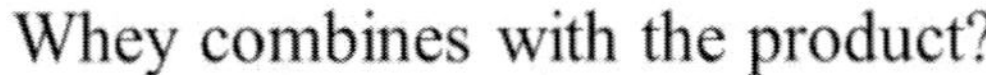

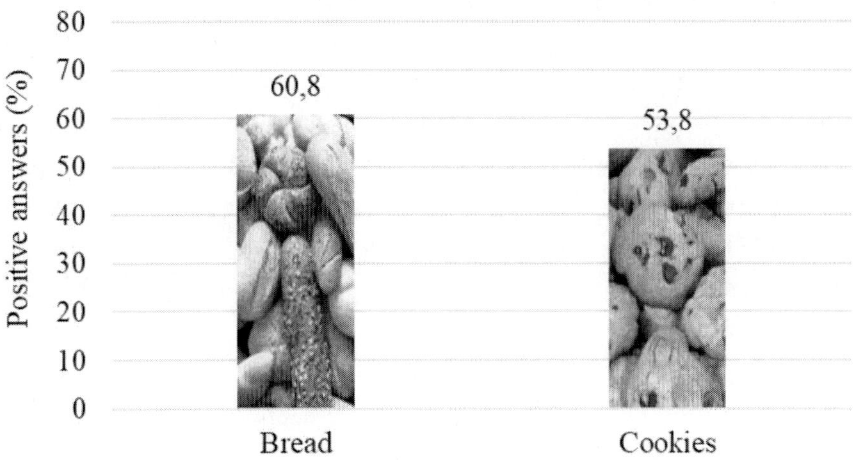

Figure 4. Opinion about the combination of whey with the bakery products.

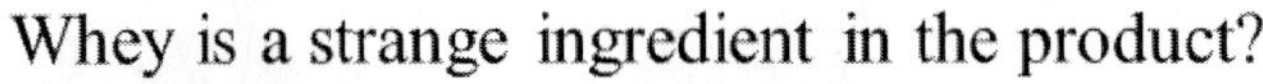

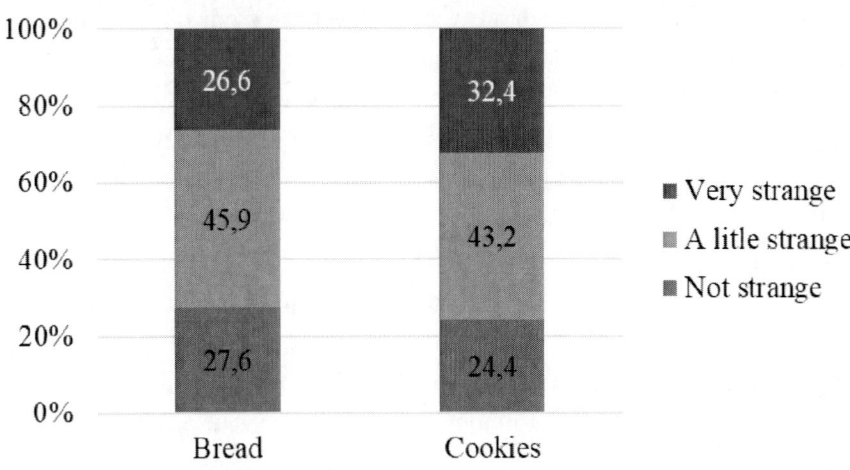

Figure 5. Opinion about how strange is the inclusion of whey into the products.

When analysing the factors influencing acceptance of baked goods incorporating whey residue, we found that only satisfaction with body weight seemed to influence product acceptance ($p = 0.019$), as shown in Table 11. Indeed, weight is a major component which influences body image and emotional attitude towards their bodies and affects people's behaviours and choices (Yarborough et al., 2016). In Spain, Carrillo et al. (2013) found that satisfaction with life positively influenced the purchase of functional foods and Schnettler et al. (2015) reported similar findings in a Chilean population.

Results presented in Figure 5 expressed the association between whey and bread or cookies. Only 27.6% and 24.4% of the responders did not consider strange the association between whey and bread or biscuits, respectively. However, it seems that the respondents considered stranger the association whey-biscuits than whey-bread.

CONCLUSION

Serra da Estrela cheese is the most well-known variety of farm cheese manufactured in central Portugal and its demand grew tremendously in recent years. Reusing the whey residues resulting from its production in baked goods such as bread and cookies can be helpful not only in reducing pollutants but also in stimulating the local economy. However, the introduction of new products to the market or the reformulation of existing ones should take into consideration the consumers' expectations and their willingness to buy such products.

Our findings are favourable to the possibility of new bakery products incorporating whey residue. Despite the fact that most respondents consider the association between whey and bread or cookies unusual, they demonstrated interest in consuming bread or cookies incorporating whey residue. Among the reasons for consuming these types of product, the most relevant factors included their high protein content, followed by environmental concerns and differences in taste. Our results also demonstrated that the people who are more satisfied with their body image have a greater willingness to purchase and consume these products.

To our knowledge, this is the first time that consumer preferences and habits regarding bread and biscuits with the addition of whey were assessed in a sample of Portuguese population. Despite some limitations of this study, such as the limited number of respondents and uneven groups of individuals according to age or gender, our results are encouraging and show that there is an opportunity to introduce in the market new baked goods incorporating whey residues from dairy production.

ACKNOWLEDGMENTS

The authors would like to acknowledge FCT - Foundation for Science and Technology, I.P., within the scope of the project Ref[a] UIDB/00681/2020. Furthermore, we would like to thank the CERNAS Research Centre and the Polytechnic Institute of Viseu for their support.

This study was supported by Instituto Politécnico de Viseu/CI&DETS through project PROJ/CI&DETS/CGD/0007 and FCT - Fundação para a Ciência e Tecnologia, I.P., under the project UID/Multi/04016/2016.

The authors thank the company who provided the whey residue: Casa da Ínsua, Viseu, Portugal, and the students from Food Quality and Nutrition for data collection: Raquel Pereira, Ana Souta, Buse Naz Gürbüz, Elisabete Almeida, Joana Lourenço, Liliana Marques, Rubina Gomes, Filipe Sousa, Carina Santos, Celeste Rocha, Christelle Marques, Cláudia Rodrigues, Filipa Manita, Márcia Félix, Sílvia Silva, Susana Rodrigues.

The authors thank the valuable advice of the experts who revised the present chapter: Professor Marijana Matek Sarić, Department of Health Studies, University of Zadar, Zadar, Croatia; Professor Monica Tarcea, Dep. of Community Nutrition & Food Safety, Univ. Medicine & Pharmacy Targu-Mures, Targu-Mures, Romania.

REFERENCES

Acquistucci, R., Melini, V., Garaguso, I. & Nobili, F. (2018). Effect of bread making process on bioactive molecules in durum wheat bread and assessment of antioxidant properties by Caco-2 cell culture model.

Journal of Cereal Science, *83*, 188–195. https://doi.org/10.1016/j.jcs.2018.08.002.

Aljobair, M. O. (2017). Assessment of the Bread Consumption Habits Among the People of Riyadh, Saudi Arabia. *Pakistan Journal of Nutrition*, *16*(5), 293–298. https://doi.org/10.3923/pjn.2017.293.298.

Aranceta, J., Pérez-Rodrigo, C., Ribas, L. & Serra-Majem, L. (2003). Sociodemographic and lifestyle determinants of food patterns in Spanish children and adolescents: The enKid study. *European Journal of Clinical Nutrition*, *57*, (Suppl 1), S40–S44. https://doi.org/10.1038/sj.ejcn.1601813.

Astrup, A., Bovy, M. W. L., Nackenhorst, K. & Popova, A. E. (2006). Food for thought or thought for food?—A stakeholder dialogue around the role of the snacking industry in addressing the obesity epidemic. *Obesity Reviews: An Official Journal of the International Association for the Study of Obesity*, *7*(3), 303–312. https://doi.org/10.1111/j.1467-789X.2006.00275.x.

Barrena-Figueroa, R. & Garcia-Lopez-de-Meneses, T. (2012). The effect of consumer innovativeness in the acceptance of a new food product. An application for the coffee market in Spain. *Spanish Journal of Agricultural Research*, *11*(3), 578–592. https://doi.org/10.5424/sjar/2013113-3903.

Bimbo, F., Bonanno, A., Nocella, G., Viscecchia, R., Nardone, G., De Devitiis, B. & Carlucci, D. (2017). Consumers' acceptance and preferences for nutrition-modified and functional dairy products: A systematic review. *Appetite*, *113*, 141–154. https://doi.org/10.1016/j.appet.2017.02.031.

Bissonnette-Maheux, V., Dumas, A. A., Provencher, V., Lapointe, A., Dugrenier, M., Straus, S., Gagnon, M. P. & Desroches, S. (2018). Women's Perceptions of Usefulness and Ease of Use of Four Healthy Eating Blog Characteristics: A Qualitative Study of 33 French-Canadian Women. *Journal of the Academy of Nutrition and Dietetics*, *118*(7), 1220-1227.e3. https://doi.org/10.1016/j.jand.2017.08.012.

Bogue, J., Collins, O. & Troy, A. J. (2017). Market analysis and concept development of functional foods. Em D. Bagchi & S. Nair (Eds.),

Developing New Functional Food and Nutraceutical Products, (pp. 29–45). Academic Press. https://doi.org/10.1016/B978-0-12-802780-6.00002-X.

Bornkessel, S., Bröring, S., (Onno) Omta, S. W. F. & van Trijp, H. (2014). What determines ingredient awareness of consumers? A study on ten functional food ingredients. *Food Quality and Preference*, *32*, 330–339. https://doi.org/10.1016/j.foodqual.2013.09.007.

Burnet, K., Kelsch, E., Zieff, G., Moore, J. B. & Stoner, L. (2019). How fitting is F.I.T.T.?: A perspective on a transition from the sole use of frequency, intensity, time, and type in exercise prescription. *Physiology & Behavior*, *199*, 33–34. https://doi.org/10.1016/ j.physbeh. 2018.11.007.

Carocho, M., Barros, L., Barreira, J. C. M., Calhelha, R. C., Soković, M., Fernández-Ruiz, V., Buelga, C. S., Morales, P. & Ferreira, I. C. F. R. (2016). Basil as functional and preserving ingredient in "Serra da Estrela" cheese. *Food Chemistry*, *207*, 51–59. https://doi.org/ 10.1016/j.foodchem.2016.03.085.

Carrillo, E., Prado-Gascó, V., Fiszman, S. & Varela, P. (2013). Why buying functional foods? Understanding spending behaviour through structural equation modelling. *Food Research International*, *50*(1), 361–368. https://doi.org/10.1016/j.foodres.2012.10.045.

Childs, J. L., Thompson, J. L., Lillard, J. S., Berry, T. K. & Drake, M. (2008). Consumer Perception of Whey and Soy Protein in Meal Replacement Products. *Journal of Sensory Studies*, *23*(3), 320–339. https://doi.org/10.1111/j.1745-459X.2008.00158.x.

Derbyshire, J. & Giovannetti, E. (2017). Understanding the failure to understand New Product Development failures: Mitigating the uncertainty associated with innovating new products by combining scenario planning and forecasting. *Technological Forecasting and Social Change*, *125*, 334–344. https://doi.org/10.1016/ j.techfore. 2017.02.007.

Frederick, D. A., Sandhu, G., Morse, P. J. & Swami, V. (2016). Correlates of appearance and weight satisfaction in a U.S. National Sample: Personality, attachment style, television viewing, self-esteem, and life satisfaction. *Body Image*, *17*, 191–203. https://doi.org/ 10.1016/j.bodyim.2016.04.001.

Gül, A., Isik, H., Bal, T. & Özer, M. (2003). Bread consumption and waste of households in urban area of Adana Province. *Electronic Journal of Polish Agricultural Universities*, *6*(2), 1–14.

Hakli, G., As¤l, E., Uçar, A., Ozdoğan, Y., Yılmaz, M., Özçelik, A., Metin, S., Sürücüo˜lu, F., Çakiro˜lu, L. & Akan, L. (2016). Nutritional Knowledge and Behavior of Adults: Their Relations with Sociodemographic Factors. *Pakistan Journal of Nutrition*, *15*(6), 532–539. https://doi.org/10.3923/pjn.2016.532.539.

Kendrick, D., Orton, E., Lafond, N., Audsley, S., Maula, A., Morris, R., Vedhara, K. & Iliffe, S. (2018). Keeping active: Maintenance of physical activity after exercise programmes for older adults. *Public Health*, *164*, 118–127. https://doi.org/10.1016/j.puhe.2018.08.003.

Kieliszek, M., Piwowarek, K., Kot, A. M., Błażejak, S., Chlebowska-Śmigiel, A. & Wolska, I. (2018). Pollen and bee bread as new health-oriented products: A review. *Trends in Food Science & Technology*, *71*, 170–180. https://doi.org/10.1016/j.tifs.2017.10.021.

Kolawole, F. L., Akinwande, B. A. & Ade-Omowaye, B. I. O. (2018). Physicochemical properties of novel cookies produced from orange-fleshed sweet potato cookies enriched with sclerotium of edible mushroom (*Pleurotus tuberregium*). *Journal of the Saudi Society of Agricultural Sciences*. https://doi.org/10.1016/j.jssas.2018.09.001.

Lagrange, V. (2015). *Most Consumers Don't Know Whey Comes from Milk.* https://blog.usdec.org/usdairyexporter/most-consumers-dont-know-whey-comes-from-milk.

Lee, S., Dangaran, K. L., Guinard, J. X. & Krochta, J. M. (2006). Consumer Acceptance of Whey-protein-coated as Compared with Shellac-coated Chocolate. *Journal of Food Science*, *67*(7), 2764–2769. https://doi.org/ 10.1111/j.1365-2621.2002.tb08812.x.

Lopes, C., Torres, D., Oliveira, A., Severo, M., Alarcão, V., Guiomar, S., Mota, J., Teixeira, P., Rodrigues, S., Lobato, L., Magalhães, V., Correia, D., Pizzarro, A., Marques, A., Vilela, S., Oliveira, L., Nicola, P., Soares, S. & Ramos, E. (2017). *Inquérito Alimentar Nacional e de Atividade*

Física, IAN-AF 2015.2016: Relatório Metodológico (Parte II; pp. 1–95). Universidade do Porto.

Lu, J. (2015). The effect of perceived carrier-ingredient fit on purchase intention of functional food moderated by nutrition knowledge and health claim. *British Food Journal*, *117*(7), 1872–1885. https://doi.org/10.1108/BFJ-11-2014-0372.

Malmqvist, T. & Wham, E. (2016). *Greendex 2014: Consumer Choice and the Environment – A Worldwide Tracking Survey - Full Report*, (pp. 1–172). National Geographic & GlobeScan. https://globescan.com/greendex-2014-consumer-choice-and-the-environment-a-worldwide-tracking-survey-full-report/.

Mannar, M. G. V. & Sankar, R. (2004). Micronutrient fortification of foods—Rationale, application and impact. *Indian Journal of Pediatrics*, *71*(11), 997–1002. https://doi.org/10.1007/bf02828115.

Mariotti, M. & Lucisano, M. (2014). Sugar and Sweeteners. Em *Bakery Products Science and Technology*, (pp. 199–221). John Wiley & Sons, Ltd. https://doi.org/10.1002/9781118792001.ch11.

Martins, N. & Ferreira, I. C. F. R. (2017). Wastes and by-products: Upcoming sources of carotenoids for biotechnological purposes and health-related applications. *Trends in Food Science & Technology*, *62*, 33–48. https://doi.org/10.1016/j.tifs.2017.01.014.

Martins, Z. E., Pinho, O. & Ferreira, I. M. P. L. V. O. (2017). Food industry by-products used as functional ingredients of bakery products. *Trends in Food Science & Technology*, *67*, 106–128. https://doi.org/10.1016/j.tifs.2017.07.003.

Menezes, E., Deliza, R., Chan, H. L. & Guinard, J. X. (2011). Preferences and attitudes towards açaí-based products among North American consumers. *Food Research International*, *44*(7), 1997–2008. https://doi.org/10.1016/j.foodres.2011.02.048.

Mirmiran, P., Mohammadi-Nasrabadi, F., Omidvar, N., Hosseini-Esfahani, F., Hamayeli-Mehrabani, H., Mehrabi, Y. & Azizi, F. (2010). Nutritional knowledge, attitude and practice of Tehranian adults and their relation to serum lipid and lipoproteins: Tehran lipid and glucose study. *Annals of Nutrition & Metabolism*, *56*(3), 233–240. https://doi.org/10.1159/000288313.

Moreira, P. A. & Padrão, P. D. (2004). Educational and economic determinants of food intake in Portuguese adults: A cross-sectional survey. *BMC Public Health*, *4*, 58. https://doi.org/10.1186/1471-2458-4-58.

Oliveira, A., Araújo, J., Severo, M., Correia, D., Ramos, E., Torres, D., Lopes, C. & by the IAN-AF Consortium. (2018). Prevalence of general and abdominal obesity in Portugal: Comprehensive results from the National Food, nutrition and physical activity survey 2015-2016. *BMC Public Health*, *18*(1), 614. https://doi.org/10.1186/s12889-018-5480-z.

Paciulli, M., Rinaldi, M., Cavazza, A., Ganino, T., Rodolfi, M., Chiancone, B. & Chiavaro, E. (2018). Effect of chestnut flour supplementation on physico-chemical properties and oxidative stability of gluten-free biscuits during storage. *LWT*, *98*, 451–457. https://doi.org/10.1016/j.lwt.2018.09.002.

Parmenter, K., Waller, J. & Wardle, J. (2000). Demographic variation in nutrition knowledge in England. *Health Education Research*, *15*(2), 163–174. https://doi.org/10.1093/her/15.2.163.

Prättälä, R., Elasoja, V. & Mykkänen, H. (2001). The consumption of rye bread and white bread as dimensions of health lifestyles in Finland. *Public Health Nutrition*, *4*(3), 813–819. https://doi.org/10.1079/PHN2000120.

Reis Lima, M. J., Fontes, L., Bahri, H., Veloso, A. C. A., Teixeira-Lemos, E. & Peres, A. M. (2019). Fatty acids profile of Serra da Estrela PDO cheeses and respective atherogenic and thrombogenic indices. *Nutrition & Food Science*, *50*(3), 417–432. https://doi.org/ 10.1108/NFS-06-2019-0178.

Rhodes, R. E., Warburton, D. E. R. & Murray, H. (2009). Characteristics of physical activity guidelines and their effect on adherence: A review of randomized trials. *Sports Medicine (Auckland, N.Z.)*, *39*(5), 355–375. https://doi.org/10.2165/00007256-200939050-00003.

Sandvik, P., Kihlberg, I., Lindroos, A. K., Marklinder, I. & Nydahl, M. (2014). Bread consumption patterns in a Swedish national dietary survey focusing particularly on whole-grain and rye bread. *Food & Nutrition Research*, *58*, (24024), 1–11. https://doi.org/10.3402/ fnr.v58.24024.

Schnettler, B., Miranda, H., Lobos, G., Sepúlveda, J., Orellana, L., Mora, M. & Grunert, K. G. (2015). Willingness to purchase functional foods according to their benefits: Consumer profiles in Southern Chile. *British Food Journal*, *117*(5), 1453–1473. https://doi.org/10.1108/BFJ-07-2014-0273.

Šereš, Z., Simović, D. Š., Grujičić, M., Maravić, N., Kiš, F., Dokić, L., Nikolić, I. & Đorđević, M. (2017, Outubro 13). *Bread as the main indicator of age-changing dietary habits among young people*. 10th International Scientific and Professional Conference "With Food to Health", Osijek, Croatia. https://doi.org/10.5281/zenodo.1161244.

Šereš, Z., Šoronja Simović, D., Grujičić, M., Maravić, N., Kiš, F., Dokić, L., Nikolić, I., Đorđević, M. & Šaranović, Ž. (2017). Bread as indicator of age-changing dietary habits among young. *Food in Health and Disease: Scientific-Professional Journal of Nutrition and Dietetics*, *6*(2), 78–84.

Tavaria, F. K., Silva Ferreira, A. C. & Malcata, F. X. (2004). Volatile Free Fatty Acids as Ripening Indicators for Serra da Estrela Cheese. *Journal of Dairy Science*, *87*(12), 4064–4072. https://doi.org/10.3168/ jds.S0022-0302(04)73548-1.

Trivedi, J. C. (2016). A Consumer Behaviour Perspective on Nutritious Biscuits. *Pacific Business Review International*, *1*(1), 66–77.

Turfani, V., Narducci, V., Durazzo, A., Galli, V. & Carcea, M. (2017). Technological, nutritional and functional properties of wheat bread enriched with lentil or carob flours. *LWT - Food Science and Technology*, *78*, 361–366. https://doi.org/10.1016/j.lwt.2016.12.030.

Valli, V., Taccari, A., Di Nunzio, M., Danesi, F. & Bordoni, A. (2018). Health benefits of ancient grains. Comparison among bread made with ancient, heritage and modern grain flours in human cultured cells. *Food Research International*, *107*, 206–215. https://doi.org/10.1016/ j.foodres.2018.02.032.

WHO. (2019, Setembro 27). *Body mass index—BMI (World health Organization).* http://www.euro.who.int/en/health-topics/disease-prevention/nutrition/a-healthy-lifestyle/body-mass-index-bmi.

Wilson, S. E. (2001). Socioeconomic status and the prevalence of health problems among married couples in late midlife. *American Journal of Public Health*, *91*(1), 131–135.

Witten, R. & Witte, J. (2009). *Statistics* (9.ª ed.). Wiley.

Yarborough, B. J. H., Leo, M. C., Yarborough, M. T., Stumbo, S., Janoff, S. L., Perrin, N. A. & Green, C. A. (2016). Improvement in Body Image, Perceived Health, and Health-Related Self-Efficacy Among People With Serious Mental Illness: The STRIDE Study. *Psychiatric Services*, *67*(3), 296–301. https://doi.org/10.1176/appi.ps.201400535.

In: Consumption and Contamination … ISBN: 978-1-53618-654-3
Editor: Egor Vagin

Chapter 5

STUDY ABOUT THE CONSUMPTION OF DAIRY PRODUCTS IN TWO COUNTRIES

Raquel P. F. Guiné*[1,*]*, Sofia G. Florença*[2]*,
***Solange Carpes*[3] *and Ofélia Anjos*[4,5]**

[1]CERNAS-IPV, Polytechnic Institute of Viseu, Viseu, Portugal
[2]Faculty of Nutrition Sciences, University of Porto, Oporto, Portugal
[3]Department of Chemistry, Federal University of Technology – Paraná (UTFPR), Pato Branco, PR, Brazil
[4]Polytechnic Institute of Castelo Branco, Castelo Branco, Portugal
[5]CEF-ISA, University of Lisbon, Lisbon, Portugal

ABSTRACT

The present chapter presents some results of a questionnaire survey undertaken on a sample of participants from Portugal and Brazil, which investigates the consumption habits of some classes of dairy products. This is a descriptive cross-sectional study made on a non-probabilistic convenience sample, consisting of 850 participants, 430 from Brazil and

* Corresponding Author's Email: raquelguine@esav.ipv.pt.

420 from Portugal. The results obtained showed that in both countries the consumption of milk and milk products as well as cheeses, butters or yogurts was very low. Nearly half or more than half of the participants never consumed these dairy products, and those who consumes them did it only once or 2-3 times per week. Small differences were identified in the dairy consumption patterns in both countries, with slightly more Portuguese participants consuming milk, cheese, butter and yogurt as compared with Brazilians.

Keywords: milk, cheese, butter, yogurt, questionnaire survey

1. Introduction

Dairy foods are considered the most basic edible products in the human food chain. Dairy products consist of a group of foods that include milk and its derivatives, like for example cheese, fermented milk and yoghurts. This group of food products has a very interesting nutritional richness, thus making it an essential group to include in dietary recommendations around the world and throughout the entire life cycle (Xu et al., 2019; Yousefi & Jafari, 2019).

In addition to providing a good amount of proteins of high biological value, dairy products also contain a high amount of vitamins and minerals, such as vitamin A, B_2, B_{12}, calcium, phosphorus, potassium and iodine. Additionally, they provide carbohydrates, especially in the form of lactose, as well as fat. In the latter case, taking into account that part of the fat from dairy products is saturated, a type of fat that has some health disadvantages, it will be important to opt for dairy products with a reduced fat content, whenever possible (Mazidi et al., 2019). Finally, dairy products also contain some substances with physiologically active properties, such as antioxidants, oligosaccharides, probiotic bacteria, butyric acid, immunoglobulins and active peptides, which have been investigated for their beneficial roles in the human body (Balthazar et al., 2018; Dantas et al., 2016; de Toledo Guimarães et al., 2018; Martins et al., 2017; Rodríguez-Pérez et al., 2019).

The consumption of dairy products has been associated with a lower incidence of many diseases, like for example cardiovascular diseases, metabolic syndromes, cancer, bone health and dental caries (Louie et al., 2013; Moosavian et al., 2020; Sonestedt et al., 2011; Tanaka et al., 2010; Zittermann, 2016). However, due to increased concern with obesity, some scientific literature has tried to associate dairy consumption with adiposity, but the observational studies on this association, performed in adults, showed inconsistent results.. While some studies have established a possible link between dairy intake and a dietary risk for obesity and some related pathologies, in other studies the consumption of dairy products has highlighted a potential protective effect against obesity. When the studies involve children or adolescents, they indicate a null or inverse association between dairy intake and obesity, resulting from the specificity of children's physical development, and, on the contrary, dairy foods, particularly milk, generally improve the health status of malnourished children (Beck et al., 2017; Dror & Allen, 2014; Oakley et al., 2010; Teegarden & Zemel, 2003; Weaver & Boushey, 2003; Xu et al., 2019).

Sociodemographic variables are usually associated with different food consumption patterns and also some behavioural factors that contribute to different food dietary trends. For example, physical activity and sedentary behaviours are associated with obesity and cardiometabolic risk, although the studies found in the scientific literature are controversial regarding the effect of dairy consumption on the development of cardiovascular disease risk factors. According to Santaliestra-Pasías et al. (2020), there is a potential protective effect of dairy consumption against cardiovascular risk factors, particularly when associated with healthy lifestyles.

The present work aims to explore the influence of some sociodemographic as well as behavioural variables on the consumption habits of dairy products in two countries, Portugal (in Europe) and Brazil (in Latin America).

2. Description of Observational Study

2.1. Instrument

This work was based on the application of a questionnaire, which was developed in order to investigate the consumption habits regarding dairy products, like milk, yogurt, cheese or butter. The questionnaire included different parts, as follows:

Part I: Sociodemographic data (5 questions);

Part II: Consuming habits regarding dairy products (21 questions distributed as follows: 4 about milk products, 7 about cheeses, 3 about butter and 7 about yogurt).

2.2. Data Collection

To undertake this descriptive cross-sectional study a non-probabilistic convenience sample was used, consisting of adult citizens from Brazil and Portugal. The survey took place through internet questionnaire and 914 answers were obtained. From those, 64 were rejected because they were incomplete or not properly filled, resulting in a sample of 850 valid responses, 430 from Brazil and 420 from Portugal. The data collection took place between November 2019 and April 2020. The answers were obtained only from adult citizens (aged 18 or over), and after informed consent. Extreme care was taken to verify all ethical guidelines when formulating and applying the questionnaire. Confidentiality of the individual answers was guaranteed to comply with ethical principles.

2.3. Data Analysis

Sociodemographic information was collected and age was classified according to: young adults (aged between 18 and 25 years), middle-aged adults (between 26 and 50 years), senior adults (between 51 and 65 years), and elderly (aged 66 years or over).

For the exploratory analysis of the data, different basic descriptive statistical tools were used. The data were processed using the SPSS program, version 26 from IBM, Inc and Excel 2016, from Microsoft Office.

3. Results and Discussion

3.1. Sociodemographic Characterization of the Sample

Table 1 shows the sociodemographic characteristics of the 850 participants included in the sample studied. The number of participants from both countries was similar, corresponding to 50.6% from Brazil and 49.4% from Portugal, but the distribution according to sex was not even, with more women (65.1%) than men (34.9%).

The participants were aged between 18 and 82 years old, classified as follows: 27.2% young adults ($18 \leq age \leq 25$ years), 47.5% middle aged adults ($26 \leq age \leq 50$ years), 23.9% senior adults ($51 \leq age \leq 65$ years) and 1.4% elderly ($age \geq 66$ years).

The distribution of the participants according to their marital status shows that most of them were married (48.5%) or single (43.8%), while 6.2% were divorced and only 1.5% were widowed.

Regarding the level of education, the large majority of the participants had completed a university degree (86.7%), and those who finished secondary school (12 school years) were less (11.9%). Participants with basic school (9 school years) and with primary school (4 school years) as the highest level of education achieved, were just a few, 0.8% and 0.6%, respectively.

Table 1. Sociodemographic characteristics of the participants

Variable	Class	N	%
Country	Brazil	430	50.6
	Portugal	420	49.4
Sex	Female	553	65.1
	Male	297	34.9
Age class	Young adults (18 ≤ age ≤ 25 years)	231	27.2
	Middle aged adults (26 ≤ age ≤ 50 years)	403	47.5
	Senior adults (51 ≤ age ≤ 65 years)	203	23.9
	Elderly (age ≥ 66 years)	12	1.4
Marital status	Single	372	43.8
	Married or living together	412	48.5
	Divorced or separated	53	6.2
	Widowed	13	1.5
Education level	Primary school (4 years)	5	0.6
	Basic school (9 years)	7	0.8
	Secondary school (12 years)	101	11.9
	University degree	737	86.7

3.2. Consuming Habits Regarding Dairy Products

3.2.1. Milk

Figure 1 shows, for both countries, the percentage of participants who never consume milk products. The percentages of people who never consume milk are very high, varying in Portugal from 47.4% for semi-kimmed milk to 94.8% for chocolate flavoured milk, and in Brazil from 46.7% for semi-skimmed milk to 85.3% for chocolate flavoured milk.

For those who consume milks, the ingestion of these products occurs with a very low frequency (Figures 2 and 3). It was observed and is shown in Figure 2 that 33.0% of Portuguese and 45.4% of Brazilians consume semi-skimmed milk seldom, revealing some differences in both countries, particularly for the frequent consumption of semi-skimmed milk, which is much more expressive among the Portuguese than the Brazilian consumers (38.5% versus 17.0%). Regarding the skimmed milk, the differences between countries are noticeable only for those who consume it seldom: 39.2% of Portuguese and 56.9% of Brazilians seldom consume skimmed milk. Also, about 20% of Portuguese always consume some type of milk (skimmed or semi-skimmed), and this percentage is similar among Brazilians, also about 20%.

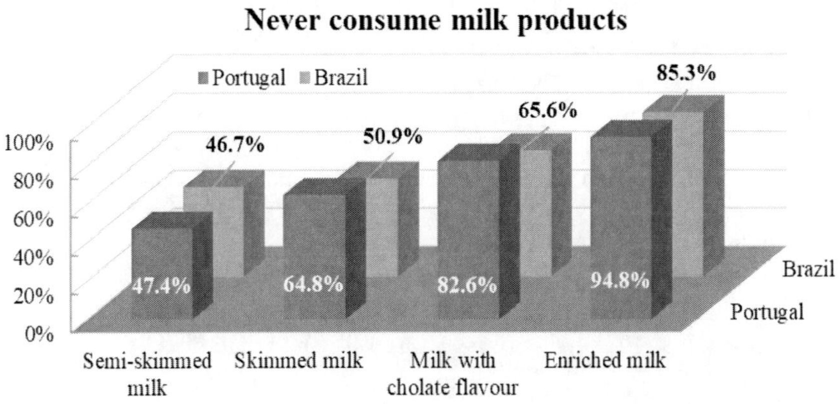

Figure 1. Percentage of participants who never consume milk products.

The results of this survey demonstrate that the consumption of milk is very low in these two countries. It has been reported in several sources that milk intake has gradually declined over the past decades. Park et al. (2019) evaluated the milk consumption patterns and perceptions in Korean participants and found that the main reasons for consuming milk by age groups were 'height growth' (30.7%) for adolescents, 'as a meal substitute' (34.8%) and 'bone health' (25.7%) for adults, and 'bone health' (59.6%) for the elderly.

According to the recommendations of the Food and Agriculture Organization of the United Nations (FAO, 2020) milk and dairy products are important components of the diet, because they provide high quality protein and micronutrients in an easily absorbed form, being appropriate for both the nutritionally vulnerable people as well as healthy people, when consumed in appropriate amounts.

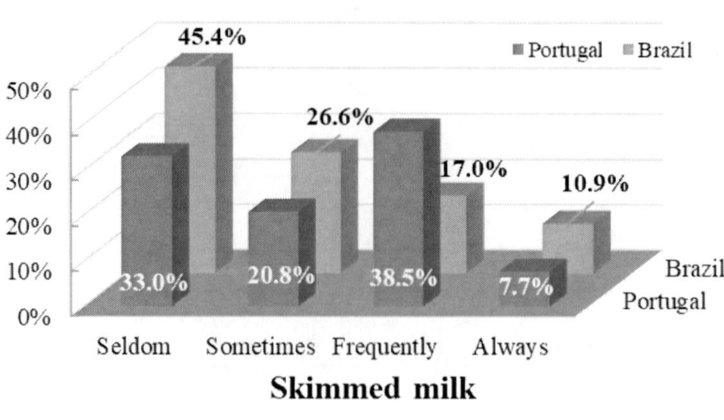

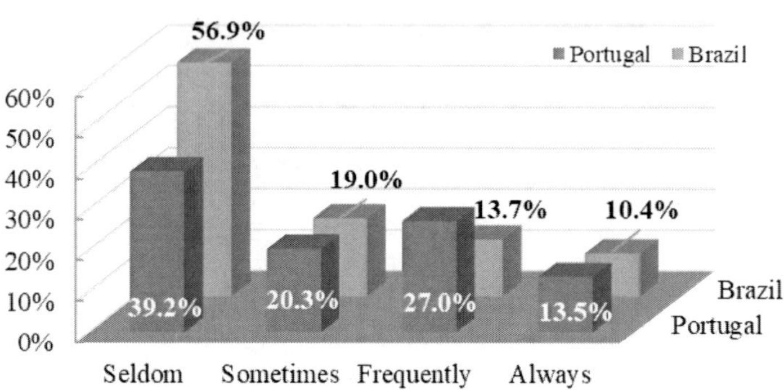

Figure 2. Frequency of consumption of semi-skimmed and skimmed milks (Legend: Seldom = once/week; Sometimes = 2-3 times/week; Frequently = once/day; Always = more than once/day).

Milk is particularly important for bone health, as many studies have confirmed (Batty & Bionaz, 2019; Cashman, 2006; Kalkwarf, 2007; Yun et al., 2020). Low milk and dairy products consumption constitutes a risk factor of osteoporosis in patients suffering from inflammatory bowel disease (Krela-Kaźmierczak et al., 2020).

Brick et al. (2020) reported that meta-analysis corroborated the protective effects of raw milk consumption in early ages on diseases such as asthma, current wheeze, hay fever or allergic rhinitis, and atopic sensitization.

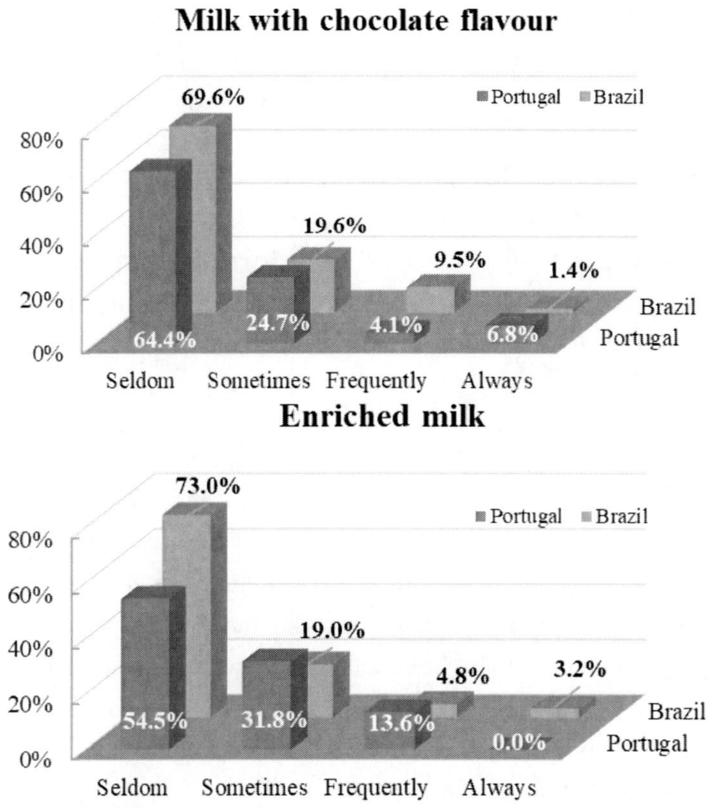

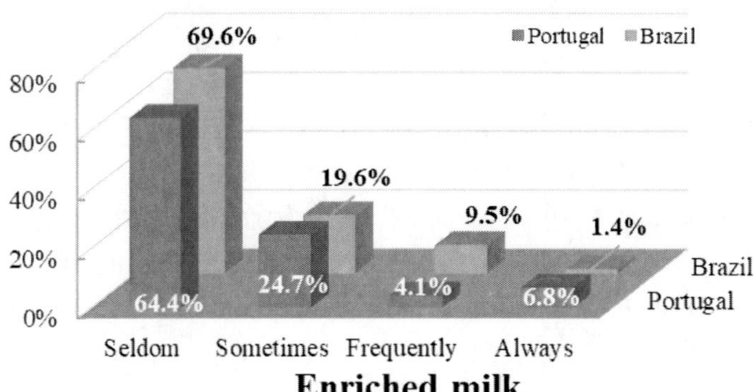

Enriched milk

Portugal Brazil

80%
60%
40%
20%
0%

73.0%
19.0%
4.8%
3.2%
54.5%
31.8%
13.6%
0.0%

Brazil
Portugal

Seldom Sometimes Frequently Always

Figure 3. Frequency of consumption of flavoured and enriched milks (Legend: Seldom = once/week; Sometimes = 2-3 times/week; Frequently = once/day; Always = more than once/day).

Many guidelines advise for regular consumption of milk and dairy products, although some studies have also demonstrated that excessive milk consumption may have some adverse effects. For example, the work by Wang et al. (2020) revealed that while moderate milk consumption diminished the risk of mortality associated with cardiovascular diseases (CVD), on the other hand, a high milk consumption showed an increased risk of cancer mortality. Another study by He et al. (2020) highlighted a positive association between milk consumption and carotid atherosclerosis in middle and old-aged Chinese people.

As for the flavoured or enriched milks (Figure 3), they are also consumed with a low frequency, 64.4% of Portuguese and 69.6% of

Brazilians seldom consume chocolate flavoured milk, being this type of product consumed always by a very low percentage of people: 6.8% for Portuguese and 1.4% for Brazilians. The enriched milks, for example with calcium and/or vitamin D, are seldom consumed by 54.5% of Portuguese and none consume this product always, while for the Brazilians 73.0% consume them seldom and 3.2% consume always.

Chocolate-flavoured milk is a popular dairy beverage in many countries around the world (Sani et al., 2019). Despite showing similar nutritional characteristics to regular milk, the high sugar content of commercialized chocolate-milks has raised concerns that culminated in their eradication from school lunch systems in the USA as well as other countries. However, chocolate-flavoured milk constitutes a nutritious alternative to other beverages, particularly for school-aged children, adolescents and young adults. It has been shown that drinking flavoured or plain milk is positive to guarantee appropriate nutrient intake in US children and adolescents, while not proven associated with adverse effects on their weight status. Still, due to high sugar contents these products have been targeted as needing to reduce added sugars, and this led to the improvement in the nutritional quality of commercial chocolate-flavoured milk by some companies (CMN, 2012; Johnson et al., 2002; Murphy et al., 2008; Oliveira et al., 2016).

New nutritional trends and nutritional demands have led to the development of a variety of enriched milks, being the most frequent those enriched with calcium and vitamin D. Calcium is an essential mineral involved in various physiological processes, such as bone health, with particular importance in childhood and adolescence for skeletal development as well as for preventing osteoporosis in menopausal women. Vitamin D is important to help the fixation of calcium in the bones, among other roles on the human body. Scientific studies have demonstrated that dietary intakes of vitamin D and calcium decrease the risk of non-communicable diseases such as obesity, diabetes mellitus, cardiovascular diseases, metabolic syndrome, and some cancers (Asbaghi et al., 2020; Erfanian et al., 2017; Meza et al., 2019; Mouratidou et al., 2013). Cow milk is among those considered suitable for enrichment with fish oil, which has shown important roles in human neurodevelopment and vascular health, due to the presence of considerable amounts of omega-3 polyunsaturated fatty acids (PUFA), such as eicosapentaenoic acid (EPA) and docosahexaenoic acid (DHA) (Costa et al., 2016; Gebreyowhans et al., 2019; Qiu et al., 2018; Uauy & Valenzuela, 2000).

3.2.2. Cheese

Cheese is a good source of calcium, fat, and protein, while also containing important amounts of vitamins A, B_2 and B_{12}, as well as other dietary minerals such as zinc or phosphorus (Ferrão & Guiné, 2019). Despite this, cheese is not a product that pleases all kinds of people due to the intense organoleptic profiles, while being appreciated by others precisely due to the richness of flavours and textures (Guiné et al., 2019).

Figure 4 shows that a high number of participants in both countries never eat cheese, being, however, this percentage higher for Brazilians as compared with the Portuguese participants. Portugal has a long tradition allied to eating cheese, highlighted by the high number (13) of traditional cheeses with PDO (Protected Designation of Origin) (Guiné et al., 2019). Despite this, the Portuguese who never eat cheese are 43.3% for semi-cured soft cheese, 45.2% for cured hard cheese, 28.8% for fresh cheese and 46.9% for whey cheese. Whey is a dilute liquid resulting from cheese manufacture that contains lactose, proteins, minerals (like calcium) and traces of fat and organic acids. The solids content is 7% (w/v) of which 75% are lactose and 10% are whey protein. Traditionally, this whey is used in whey cheese production, such as "Requeijão" in Portugal, particularly made with whey from sheep milk. This Portuguese whey cheese has been reported as a food vector for beneficial conditions favouring the gastrointestinal system (Gomes da Cruz et al., 2009; Guiné et al., 2012).

Regarding the sliced cheeses, 35.0% never eat semi-fat cheese and 56.7% never eat low fat cheese. The corresponding percentages for Brazilians are always higher as compared with the Portuguese, except for whey cheese, which seems to be more frequent among Brazilians. In Brazil several types of cheeses integrate Brazilian culture and tradition. Although there are some typically Brazilian cheeses, there are others which have been inspired by the knowledge brought to the country by foreigners. In general, the original versions of those foreign cheeses were adapted to the Brazilian milk supply and conditions in the different dairy basins and also to the preferences of Brazilian consumers (Ferreira et al., 2019). One of the key features of the Brazilian cheese market consists in the different definitions and classifications adopted to distinguish the product. They vary according to place of origin, type of milk (cow, sheep, goat, buffalo), manufacturing procedures, texture, maturation time, among other factors (Ferreira et al., 2019).

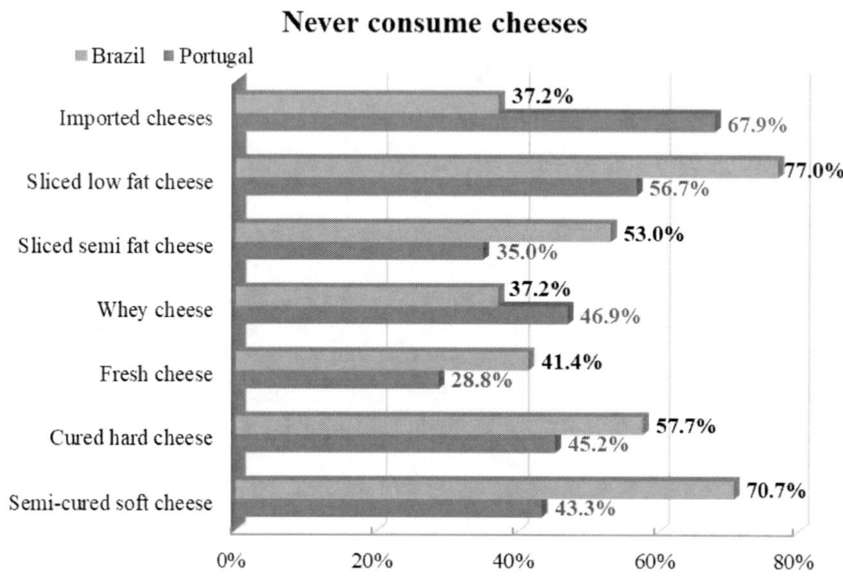

Figure 4. Percentage of participants who never consume specific types of cheese.

Portugal has good pastures, and pastoralism is a constant and very ancient activity at countryside. This traditional activity provides a means of subsistence for rural families, coming from the cattle or the derived products, such as dairy, including cheese. There are several types of cheese in Portugal, made with cow, goat, sheep or mixture milks, and with different tastes and consistencies. The consistency of the paste, the taste and the degree of fat and protein are also variable. There are several types of Portuguese traditional cheeses, which can be classified according to: (i) the type of milk used for the cheese production; (ii) the fat content; (iii) the type and intensity of ripening or (iv) the consistency of the paste (Guiné et al., 2019). This diversity and richness of the Portuguese cheeses against the Brazilian can help explaining the differences observed between the participants form these two countries regarding the consumption of imported cheeses. While in Portugal almost 70% never eat imported cheeses, in Brazil that number is reduced to 37.2% (Figure 4).

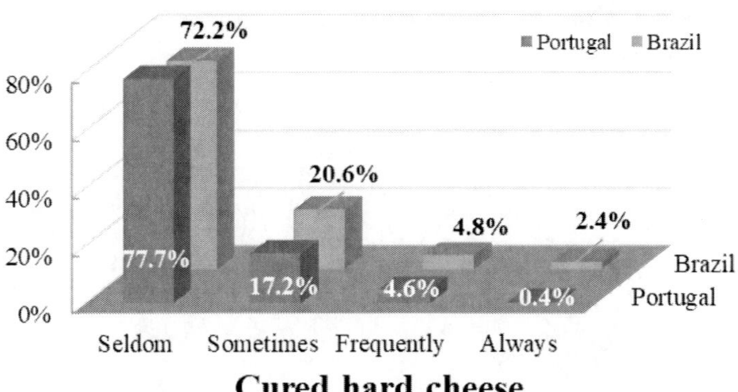

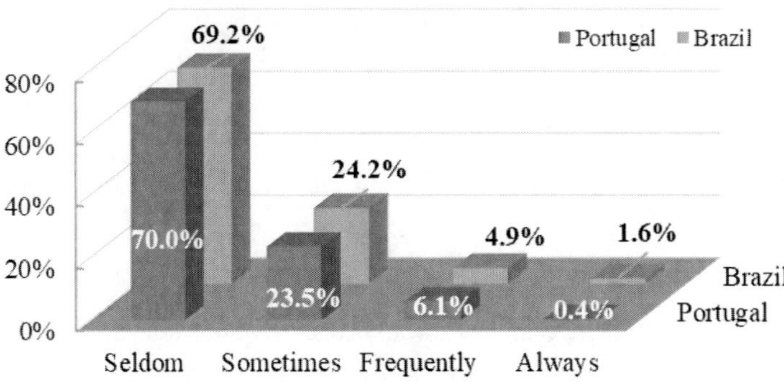

Figure 5. Frequency of consumption of semi-cured and cured cheeses (Legend: Seldom = once/week; Sometimes = 2-3 times/week; Frequently = once/day; Always = more than once/day).

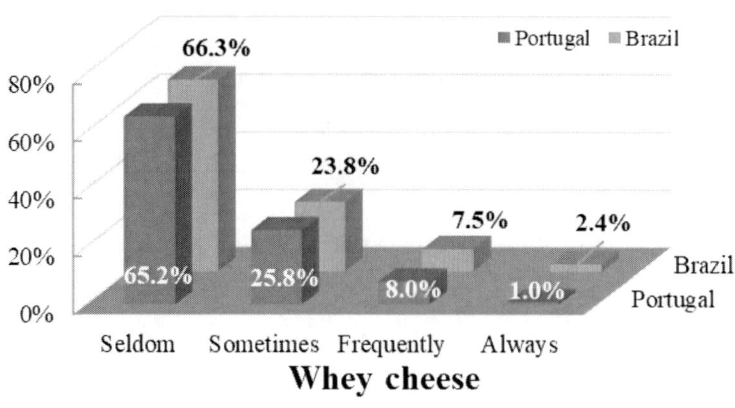

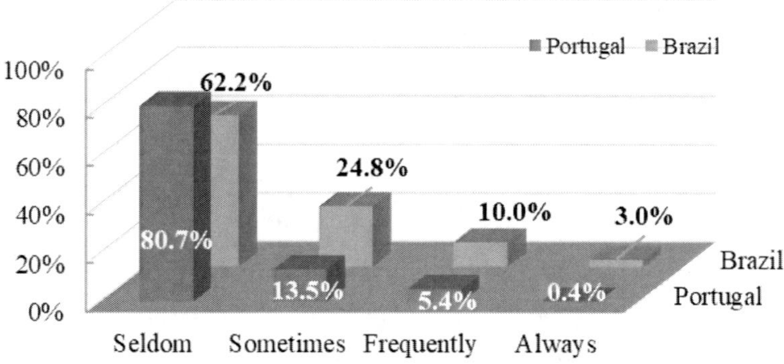

Figure 6. Frequency of consumption of fresh and whey cheese (Legend: Seldom = once/week; Sometimes = 2-3 times/week; Frequently = once/day; Always = more than once/day).

For the participants who consumed cheese, the frequency of consumption of different types of cheese was investigated and is shown in the following Figures (5 to 7). Figure 5 presents the frequency of consumption of semi-cured and cured cheeses in both countries, and it can be observed that in both cases (semi-cured soft cheese and cured hard cheese) there are similar trends for consumption among Portuguese and Brazilian consumers. These two types of cheese are consumed with a low frequency with a very expressive percentage consuming them only sporadically, and regular consumption is present only for about 5% of consumers. The semi-cured soft cheese consumption is slightly higher among Brazilians as compared with Portuguese (lower percentages for seldom and higher for the other options). A similar trend was also observed for cured hard cheese.

Figure 6 shows the frequency of consumption of fresh and whey cheeses in Portugal and Brazil. The results indicated that fresh cheese consumption is also very similar in both countries, with a very high percentage of participants seldom consuming it (65.2% of Portuguese and 66.3% of Brazilians).

Those who sometimes consume fresh cheese, i.e., 2-3 times/week, are slightly over 20% (25.8% and 23.8%, respectively for Portuguese and Brazilian consumers) (Figure 6). This is similar to the values found for the semi-cured and cures types of cheese (Figure 5). About 10% of Portuguese and also Brazilians consume fresh cheese regularly (frequently or always).

The consumption of whey cheese is lower for Portuguese as compared with Brazilian consumers: 80.7% of the Portuguese eat it seldom while that percentage for Brazilians is lower, 62.2%. On the other hand, more Brazilians eat whey cheese in higher percentages for more frequent options: 24.8% eat it sometimes against only 13.5% of Portuguese, 10.0% eat it frequently, on a daily basis, against 5.4% of Portuguese and 3.0% eat it always, i.e., more than once/day against 0.4% of the Portuguese.

Figure 7 presents the frequency of consumption of sliced cheeses. This sliced portioned and ready to eat alternative, is a convenient form, that intends to facilitate and incentive cheese consumption. These alternatives seem more popular among the Portuguese than the Brazilians: 47.6% and 58.8% seldom consume sliced semi-fat and low-fat cheese, respectively, while the number of Brazilian participants who seldom consume them is considerably high: 69.8% and 82.8%.

For the same types of sliced cheese, the percentage of Portuguese who consume them sometimes, frequently and always is always higher than for Brazilians. Comparing both options, i.e., the semi-fat sliced cheese with the low-fat version, this last is less consumed and with a lower frequency in both countries (Figures 7 and 4). These results are surprising having in mind the modern consumer trends. It has been reported that consumption of low-fat foods is increasing every year, since the consumption of excessive fat may impact health, by increasing the risk of obesity, cancer, and cardiovascular diseases (Beck et al., 2017; Dror & Allen, 2014; Oakley et al., 2010; Xu et al., 2019). Cheese is not only consumed in its original form, and during the last decade, it has become one of the most widely used food ingredients, leading to the development of several types of low-fat cheeses that have health-promoting benefits beyond their nutritional value (Oluk et al., 2014; Wang et al., 2019).

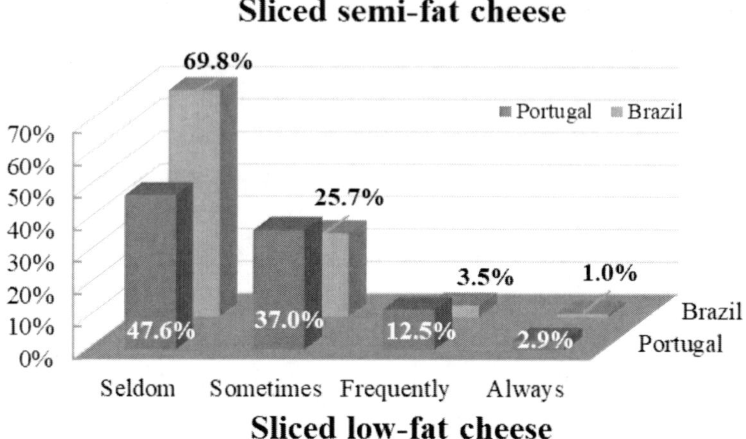

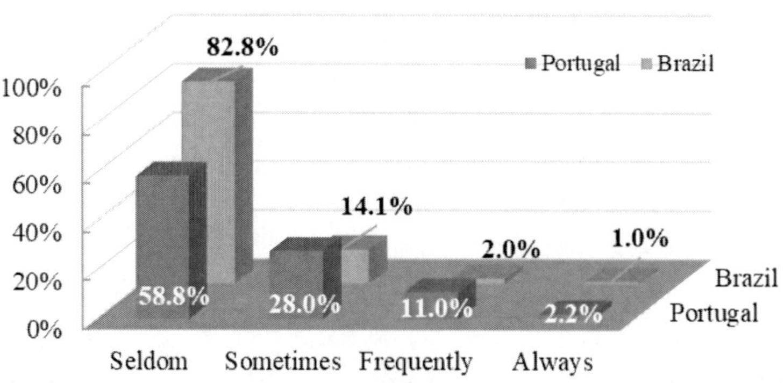

Figure 7. Frequency of consumption of sliced cheeses (Legend: Seldom = once/week; Sometimes = 2-3 times/week; Frequently = once/day; Always = more than once/day).

Figure 8 shows the frequency of consumption of imported cheeses in both countries. As previously highlighted, the Portuguese seldom eat these products (nearly 90%), while about 1/4 of Brazilians consume them sometimes (2-3 times/week), 10% eat them on a daily basis and 3% more than once per day.

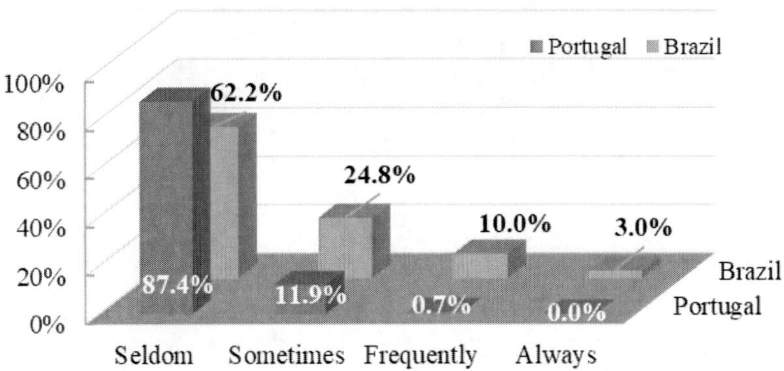

Figure 8. Frequency of consumption of imported cheeses (Legend: Seldom = once/week; Sometimes = 2-3 times/week; Frequently = once/day; Always = more than once/day).

3.2.3. Butter

Butter is reported as one of the most ancient and popular dairy products, being quite expensive and with an important economic value in the dairy industry. Butter contains valuable fatty acids with documented help benefits, as well as fat-soluble vitamins (A, D, E, K), tocopherols and carotenoids, among its important nutrients. Therefore, butter has nutritional and functional properties besides economic value in the dairy industry (Fadzillah et al., 2015; Nurrulhidayah et al., 2015; Taylan et al., 2020).

Butter is manufactured by transforming dairy cream, which is an oil-in-water emulsion, into a water-in-oil emulsion by means of a mechanical phase-inversion process. Hence, the final product microstructure consists of a continuous oil phase holding aggregates of fat in the crystalized form as well as intact and damaged fat globules, in which water and air droplets are dispersed. Present international standards establish the maximum water content and minimum fat content in butter as 16% and 80%, respectively (IFS, 2011; Truong et al., 2018).

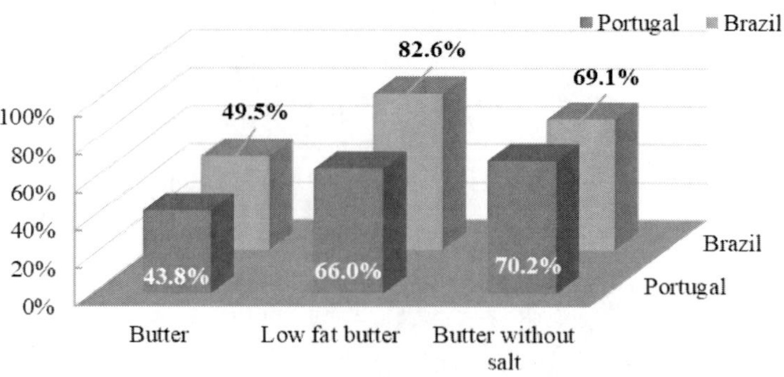

Figure 9. Percentage of participants who never consume specific types of butter.

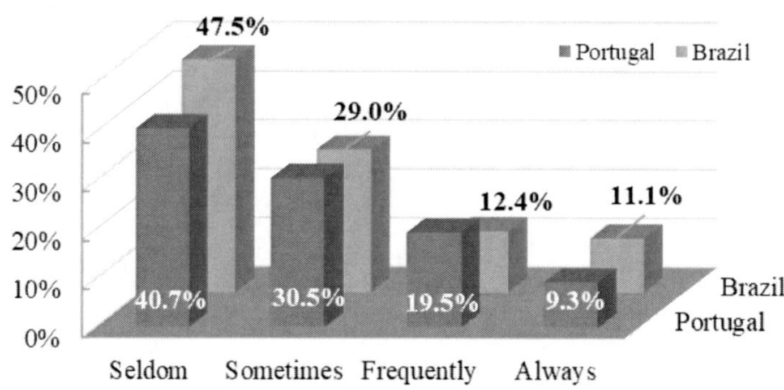

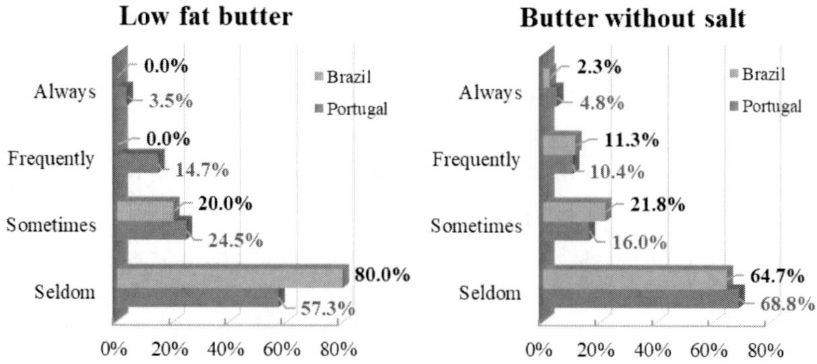

Figure 10. Frequency of consumption of butter (Legend: Seldom = once/week; Sometimes = 2-3 times/week; Frequently = once/day; Always = more than once/day).

Figure 9 represents the percentage of participants who never consume butter in Portugal and Brazil. Around 50% never consume regular butter, but these percentages increase for the consumption of low-fat and no salt butter. Low fat is never consumed by 66.0% of the Portuguese and by 82.6% of the Brazilians, while no salt butter is never consumed by 70.2% of Portuguese and by 69.1% of Brazilians (Figure 9).

Figure 10 shows the frequency of consumption of butter among the Portuguese and Brazilian participants. From those who consume butter, mostly they consume regular butter only sporadically or sometimes (about 30% consume it 2-3 times/week, in both countries).

Low-fat and no salt butters are consumed very sporadically in both countries. Only 24.5% of Portuguese and 20.0% Brazilians consume low-fat butter sometimes, and only 16.0% of Portuguese and 21.8% Brazilians consume sometimes the no salt butter (Figure 10).

3.2.4. Yogurt

Yogurt is included in the category of fermented milk products. These are dairy products that have been fermented by lactic acid bacteria such as *Lactobacillus*, *Lactococcus*, and *Leuconostoc*. This fermentation process allows extension of the product shelf life while enhancing its taste and improving its digestibility as compared to milk. Fermented dairy products' consumption has been increasing widely around the world and market trends indicate that this increase will possibly continue. This is justified because there is a growing consumer interest in these products owing not only to the nutritional value but also to the health benefits they offer, most especially considering that effect on the intestinal bacterial microbiota contributes to a healthy life and to increase life expectancy (Bourrie et al., 2016; Chen et al., 2019; García-Burgos et al., 2020). Yogurts contain large quantities of live bacteria, which have benefits for human health, such as contributing to the maintenance and balance of the intestinal flora, facilitating digestion and preventing constipation and other gastrointestinal disorders. Additionally, several studies have reported that yogurt and other fermented dairy products have antimutagenic and anticarcinogenic effects, besides providing protection against colorectal adenomas (Guiné et al., 2016).

The results of the present study revealed that a high percentage of participants never consume yogurt (Figure 11): varying from 33.1% to 73.3% in Portugal, depending on the type of yogurt, and from 52.3% to 74.2% in Brazil. This result is contradictory to that reported by Guiné (2019) according to which 73.1% of Portuguese consumers eat yogurts regularly, and most especially women. Those yogurts which are never consumed by a higher percentage of participants are the bicompartmented ones, which separate the yogurt form cereals or fruit or chocolate chips, and which are particularly addressed for children. On the other hand, the liquid yogurts are those that are not consumed by a low percentage of participants: 33.1% among Portuguese and 52.3% among Brazilians.

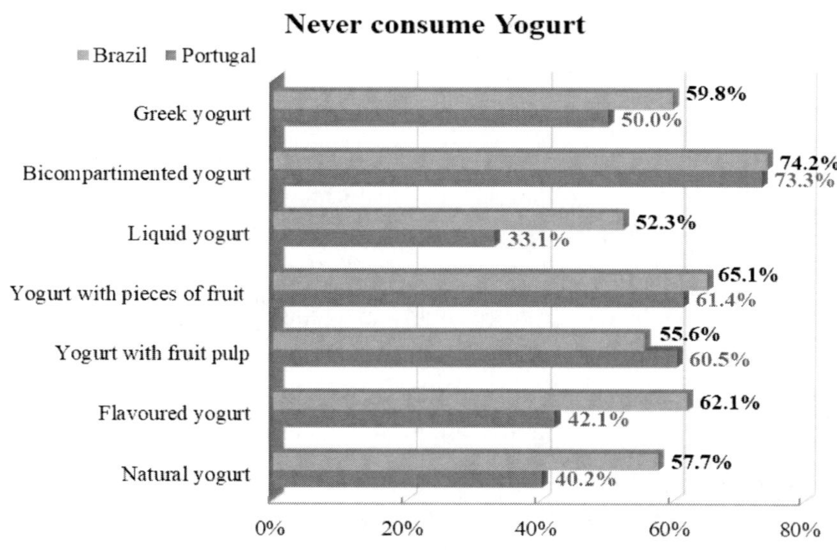

Figure 11. Percentage of participants who never consume specific types of yogurt.

Figure 12 shows the frequency of consumption for natural and flavoured types of yogurt. The natural yogurt is consumed more frequently by the Portuguese: 32.3% eat them sometimes, 2-3 times/week, against 18.7% of Brazilians and 14.7% eat them on a daily basis while that percentage is only 5.5% for Brazilians. A similar trend is observed also for the flavoured yogurt, although with a slightly lower consumption as compared with the natural yogurt (Figures 11 and 12).

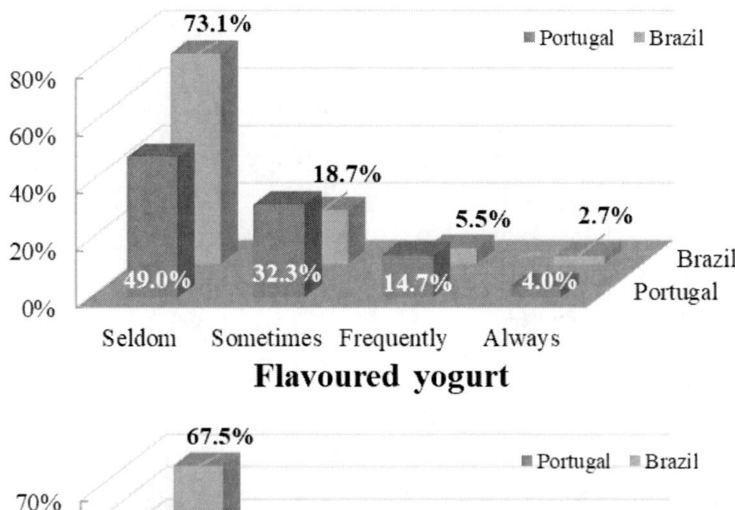

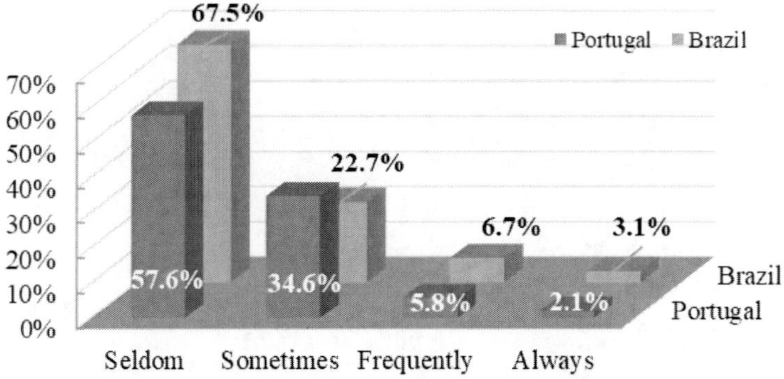

Figure 12. Frequency of consumption of natural and flavoured yogurts (Legend: Seldom = once/week; Sometimes = 2-3 times/week; Frequently = once/day; Always = more than once/day).

According to the study by Guiné (2019), about 50% of the Portuguese consume yogurts on a daily basis, and about 40% consume them 2-3 times/week, thus indicating that yogurt was a staple food in the daily diet of the Portuguese. Possible differences might be due to the external factors associated with these two studies, since the present study was undertaken under special circumstances related to COVID-19 pandemic during the first part of the year 2020. In the same study it was reported that flavoured yogurts were the most consumed (61.3%) if compared with other types of yogurt.

Figure 13 presents the frequency of consumption of yogurts with fruit pulp and fruit pieces. The results show that these two products have a very similar consumption trend in both countries, maybe just highlighting that the yogurts with pieces of fruit were a little less consumed by Brazilian than by Portuguese consumers. The study by Guiné (2019) identified these two types of yogurts as being less important in the market of yogurts in Portugal.

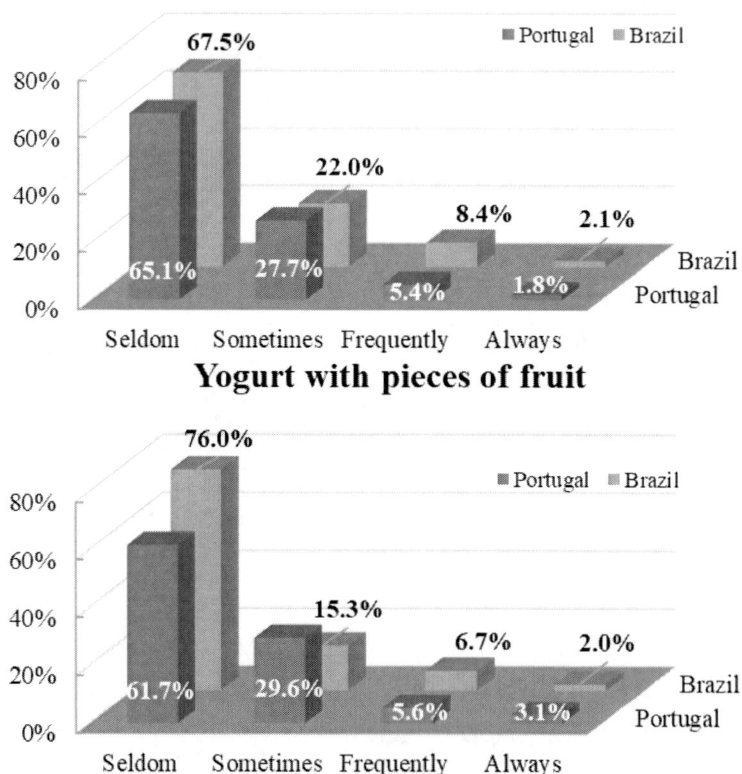

Figure 13. Frequency of consumption of fruit yogurts (Legend: Seldom = once/week; Sometimes = 2-3 times/week; Frequently = once/day; Always = more than once/day).

The results in Figure 14 indicate that liquid yogurts are more popular among Portuguese consumers than Brazilians. While 71.2% of Brazilians consume them seldom and only 20.0% consume them sometimes, for the

Portuguese, 48.0% consume them seldom, 30.6% sometimes and 16.7% once/day and almost 5% more than once per day.

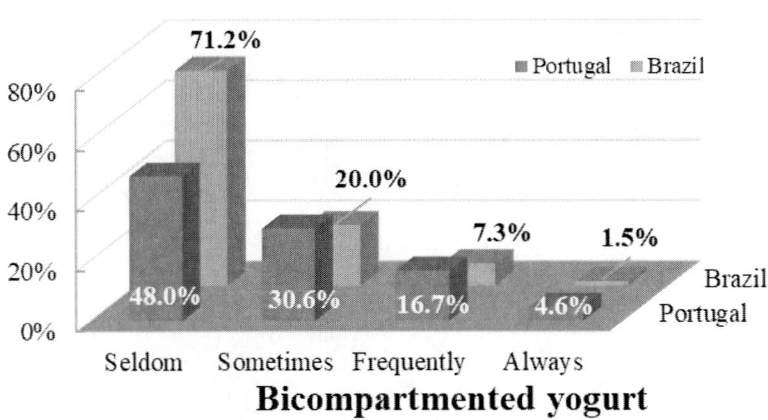

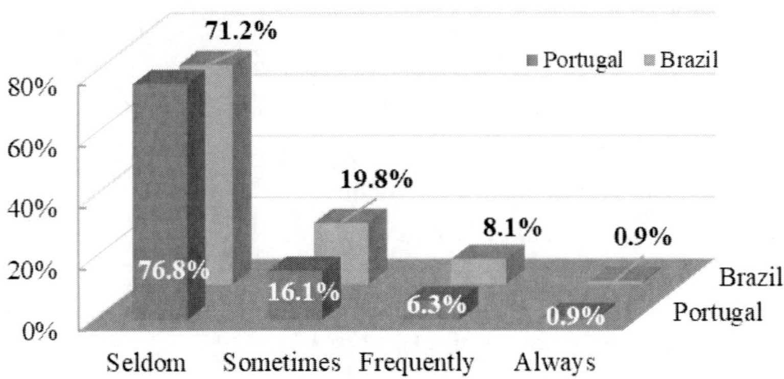

Figure 14. Frequency of consumption of liquid and bicompartmented yogurt (Legend: Seldom = once/week; Sometimes = 2-3 times/week; Frequently = once/day; Always = more than once/day).

Figure 14 also shows that the consumption of bicompartmented yogurts is just sporadic in Portugal as well as in Brazil, thus corroborating what had previously been observed for this type of product.

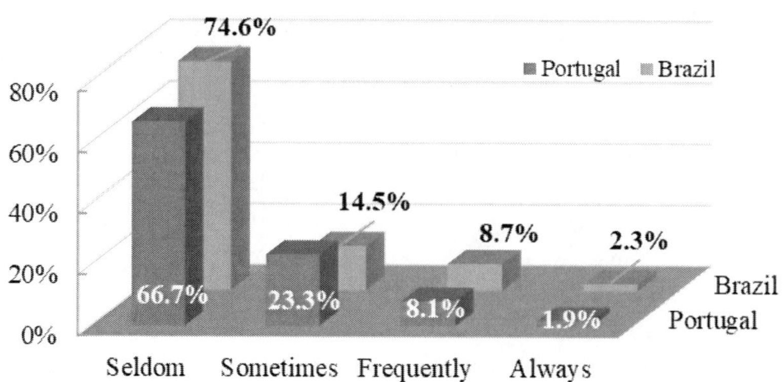

Figure 15. Frequency of consumption of Greek yogurt (Legend: Seldom = once/week; Sometimes = 2-3 times/week; Frequently = once/day; Always = more than once/day).

The Greek type yogurts have become increasingly popular in many countries around the world. However, in this study it was observed that both Portuguese and Brazilian consumers seldom eat this type of product (Figure 15), 66.7% and 74.6%, respectively. Greek yogurt is a concentrated fermented milk product, with a high protein content (up to 10%) and low or no fat, the consistency is much appreciated by consumers. It presents a creamy and thick consistency, without the need of stabilizers (Brown & Chambers, 2015; Desai et al., 2013; Fekete et al., 2013; Körzendörfer et al., 2019; Tamime et al., 2014).

CONCLUSION

The results of the present survey allowed identifying the consumption patterns regarding some types of dairy products among participants (consumers) from Portugal and Brazil. The obtained results indicated that, in both countries, the consumption of milk and milk products is very low, and similar trends were found for cheeses, butter and yogurts. Nearly half or more than half of the participants never consume these dairy products, and those who consume them do it seldom or just a few times per week. Small differences were encountered between both countries' habits, with slightly more Portuguese participants consuming milk, cheese, butter and yogurt as compared with Brazilians.

ACKNOWLEDGMENTS

The authors would like to acknowledge FCT - Foundation for Science and Technology, I.P., within the scope of the project Refª UIDB/00681/2020. Furthermore, we would like to thank the CERNAS Research Centre and the Polytechnic Institute of Viseu for their support.

Thanks to Centro de Estudos Florestais is a research unit funded by FCT - Foundation for Science and Technology, under the Project (UIDB/00239/2020).

The authors thank the valuable advice of the experts who revised the present chapter: Professor Maria João Barroca from Agrarian School of Polytechnic Institute of Coimbra, Portugal; Professor João Carlos Gonçalves from Agrarian School of Polytechnic Institute of Viseu, Portugal.

REFERENCES

Asbaghi, O., Sadeghian, M., Mozaffari-Khosravi, H., Maleki, V., Shokri, A., Hajizadeh-Sharafabad, F., Alizadeh, M., & Sadeghi, O. (2020). The effect of vitamin d-calcium co-supplementation on inflammatory biomarkers: A systematic review and meta-analysis of randomized controlled trials. *Cytokine*, *129*, 155050. https://doi.org/10.1016/j.cyto.2020.155050.

Balthazar, C. F., Silva, H. L. A., Esmerino, E. A., Rocha, R. S., Moraes, J., Carmo, M. A. V., Azevedo, L., Camps, I., K.D Abud, Y., Sant'Anna, C., Franco, R. M., Freitas, M. Q., Silva, M. C., Raices, R. S. L., Escher, G. B., Granato, D., Senaka Ranadheera, C., Nazarro, F., & Cruz, A. G. (2018). The addition of inulin and Lactobacillus casei 01 in sheep milk ice cream. *Food Chemistry*, *246*, 464–472. https://doi.org/10.1016/j.foodchem.2017.12.002.

Batty, B. S., & Bionaz, M. (2019). Graduate Student Literature Review: The milk behind the mustache: A review of milk and bone biology. *Journal of Dairy Science*, *102*(8), 7608–7617. https://doi.org/10.3168/jds.2019-16421.

Beck, A. L., Heyman, M., Chao, C., & Wojcicki, J. (2017). Full fat milk consumption protects against severe childhood obesity in Latinos. *Preventive Medicine Reports*, *8*, 1–5. https://doi.org/10.1016/j.pmedr.2017.07.005.

Bourrie, B. C. T., Willing, B. P., & Cotter, P. D. (2016). The Microbiota and Health Promoting Characteristics of the Fermented Beverage Kefir. *Frontiers in Microbiology*, *7*. https://doi.org/10.3389/fmicb.2016.00647.

Brick, T., Hettinga, K., Kirchner, B., Pfaffl, M. W., & Ege, M. J. (2020). The Beneficial Effect of Farm Milk Consumption on Asthma, Allergies, and Infections: From Meta-Analysis of Evidence to Clinical Trial. *The Journal of Allergy and Clinical Immunology: In Practice*, *8*(3), 878-889.e3. https://doi.org/10.1016/j.jaip.2019.11.017.

Brown, M. D., & Chambers, D. H. (2015). Sensory Characteristics and Comparison of Commercial Plain Yogurts and 2 New Production Sample Options. *Journal of Food Science*, *80*(12), S2957-2969. https://doi.org/10.1111/1750-3841.13128.

Cashman, K. D. (2006). Milk minerals (including trace elements) and bone health. *International Dairy Journal*, *16*(11), 1389–1398. https://doi.org/10.1016/j.idairyj.2006.06.017.

Chen, M., Ye, X., Shen, D., & Ma, C. (2019). Modulatory Effects of Gut Microbiota on Constipation: The Commercial Beverage Yakult Shapes Stool Consistency. *Journal of Neurogastroenterology and Motility*, *25*(3), 475–477. https://doi.org/10.5056/jnm19048.

CMN. (2012). Industry continues to push for more school milk options. *Cheese Market News*, *31*, 1–12.

Costa, M., Losada-Barreiro, S., Paiva-Martins, F., & Bravo-Díaz, C. (2016). Optimizing the efficiency of antioxidants in emulsions by lipophilization: Tuning interfacial concentrations. *RSC Advances*, *6*(94), 91483–91493. https://doi.org/10.1039/C6RA18282H.

Dantas, A. B., Jesus, V. F., Silva, R., Almada, C. N., Esmerino, E. A., Cappato, L. P., Silva, M. C., Raices, R. S. L., Cavalcanti, R. N., Carvalho, C. C., Sant'Ana, A. S., Bolini, H. M. A., Freitas, M. Q., & Cruz, A. G. (2016). Manufacture of probiotic Minas Frescal cheese with Lactobacillus casei Zhang. *Journal of Dairy Science*, *99*(1), 18–30. https://doi.org/10.3168/jds.2015-9880.

de Toledo Guimarães, J., Silva, E. K., de Freitas, M. Q., de Almeida Meireles, M. A., & da Cruz, A. G. (2018). Non-thermal emerging technologies and their effects on the functional properties of dairy products. *Current Opinion in Food Science*, *22*, 62–66. https://doi.org/10.1016/j.cofs.2018.01.015.

Desai, N. T., Shepard, L., & Drake, M. A. (2013). Sensory properties and drivers of liking for Greek yogurts. *Journal of Dairy Science*, *96*(12), 7454–7466. https://doi.org/10.3168/jds.2013-6973.

Dror, D. K., & Allen, L. H. (2014). Dairy product intake in children and adolescents in developed countries: Trends, nutritional contribution, and a review of association with health outcomes. *Nutrition Reviews*, *72*(2), 68–81. https://doi.org/10.1111/nure.12078.

Erfanian, A., Rasti, B., & Manap, Y. (2017). Comparing the calcium bioavailability from two types of nano-sized enriched milk using in-vivo assay. *Food Chemistry*, *214*, 606–613. https://doi.org/10.1016/j.foodchem.2016.07.116.

Fadzillah, N. A., Man, Y. bin C., Rohman, A., Rosman, A. S., Ismail, A., Mustafa, S., & Khatib, A. (2015). Detection of Butter Adulteration with Lard by Employing (1)H-NMR Spectroscopy and Multivariate Data Analysis. *Journal of Oleo Science*, *64*(7), 697–703. https://doi.org/10.5650/jos.ess14255.

FAO. (2020). *Milk and Dairy Products in Human Nutrition—Questions and Answers* (pp. 1–4). Food and Agriculture Organization of the United Nations.

Fekete, Á. A., Givens, D. I., & Lovegrove, J. A. (2013). The impact of milk proteins and peptides on blood pressure and vascular function: A review of evidence from human intervention studies. *Nutrition Research Reviews*, *26*(2), 177–190. https://doi.org/10.1017/S0954422413000139.

Ferrão, A., & Guiné, R. (2019). Cheese: Nutritional Aspects and Health Effects. Em *Cheeses around the World: Types, Production, Properties and Cultural and Nutritional Relevance* (pp. 17–46). Nova Science Publishers.

Ferreira, V., Pires, I. S. C., Miranda, L. S., & Ribeiro, M. C. (2019). Brazilian Cheeses: Diversity, Culture and Tradition. Em *Cheeses around the World: Types, Production, Properties and Cultural and Nutritional Relevance* (pp. 17–46). Nova Science Publishers.

García-Burgos, M., Moreno-Fernández, J., Alférez, M. J. M., Díaz-Castro, J., & López-Aliaga, I. (2020). New perspectives in fermented dairy products and their health relevance. *Journal of Functional Foods*, *72*, 104059. https://doi.org/10.1016/j.jff.2020.104059.

Gebreyowhans, S., Lu, J., Zhang, S., Pang, X., & Lv, J. (2019). Dietary enrichment of milk and dairy products with n-3 fatty acids: A review. *International Dairy Journal*, *97*, 158–166. https://doi.org/10.1016/j.idairyj.2019.05.011.

Gomes da Cruz, A., Alonso Buriti, F. C., Batista de Souza, C. H., Fonseca Faria, J. A., & Isay Saad, S. M. (2009). Probiotic cheese: Health benefits, technological and stability aspects. *Trends in Food Science & Technology*, *20*(8), 344–354. https://doi.org/10.1016/j.tifs.2009.05.001.

Guiné, R. de P. F. (2019). Study of Consumer Acceptance by Means of Questionnaire Survey Towards Newly Developed Yogurts with Functional Ingredients. *Current Nutrition & Food Science*, *15*(3), 243–256.

Guiné, R. P. F., Costa, E., Santos, S., Correia, A. C., Correia, P. M. R., & Pato, L. (2012). Food Product development: Whey cheese with pumpkin jam. *Academic Research International*, *2*(1), 52–59.

Guiné, R. P. F., Florença, S. G., & Correia, P. M. R. (2019). Portuguese Traditional Cheeses: Production and Characterization. Em *Cheeses around the World: Types, Production, Properties and Cultural and Nutritional Relevance* (pp. 115–161). Nova Science Publishers.

Guiné, Raquel P. F., Rodrigues, A. P., Ferreira, S. M., & Gonçalves, F. J. (2016). Development of Yogurts Enriched with Antioxidants from Wine. *Journal of Culinary Science & Technology*, *14*(3), 263–275. https://doi.org/10.1080/15428052.2015.1111180.

He, M., Guo, Z., Lu, Z., Wei, S., & Wang, Z. (2020). High milk consumption is associated with carotid atherosclerosis in middle and old-aged Chinese. *International Journal of Cardiology Hypertension*, *5*, 100031. https://doi.org/10.1016/j.ijchy.2020.100031.

IFS. (2011). *Codex Alimentarius: Milk and Milk Products* (Codex Standard 279-1971). Food and Agriculture Organization of the United Nations.

Johnson, R., Frary, C., & Wang, M. (2002). The nutritional consequences of flavored-milk consumption by school-aged children and adolescents in the United States. *Journal of the American Dietetic Association*, *102*, 853–856. https://doi.org/10.1016/S0002-8223(02)90192-6.

Kalkwarf, H. J. (2007). Childhood and adolescent milk intake and adult bone health. *International Congress Series*, *1297*, 39–49. https://doi.org/10.1016/j.ics.2006.08.008.

Körzendörfer, A., Schäfer, J., Hinrichs, J., & Nöbel, S. (2019). Power ultrasound as a tool to improve the processability of protein-enriched fermented milk gels for Greek yogurt manufacture. *Journal of Dairy Science*, *102*(9), 7826–7837. https://doi.org/10.3168/jds.2019-16541.

Krela-Kaźmierczak, I., Michalak, M., Szymczak-Tomczak, A., Czarnywojtek, A., Wawrzyniak, A., Łykowska-Szuber, L., Stawczyk-Eder, K., Dobrowolska, A., & Eder, P. (2020). Milk and dairy product consumption in patients with inflammatory bowel disease: Helpful or harmful to bone mineral density? *Nutrition*, *79–80*, 110830. https://doi.org/10.1016/j.nut.2020.110830.

Louie, J. C. Y., Flood, V. M., Rangan, A. M., Burlutsky, G., Gill, T. P., Gopinath, B., & Mitchell, P. (2013). Higher regular fat dairy consumption is associated with lower incidence of metabolic syndrome but not type 2 diabetes. *Nutrition, Metabolism, and Cardiovascular Diseases: NMCD*, *23*(9), 816–821. https://doi.org/10.1016/j.numecd.2012.08.004.

Martins, N., Oliveira, M. B. P. P., & Ferreira, I. C. F. R. (2017). Development of Functional Dairy Foods. Em J.-M. Mérillon & K. G. Ramawat (Eds.), *Bioactive Molecules in Food* (pp. 1–19). Springer International Publishing. https://doi.org/10.1007/978-3-319-54528-8-5-1.

Mazidi, M., Mikhailidis, D. P., Sattar, N., Howard, G., Graham, I., & Banach, M. (2019). Consumption of dairy product and its association with total and cause specific mortality – A population-based cohort study and meta-analysis. *Clinical Nutrition*, *38*(6), 2833–2845. https://doi.org/10.1016/j.clnu.2018.12.015.

Meza, B. E., Zorrilla, S. E., & Olivares, M. L. (2019). Rheological methods to analyse the thermal aggregation of calcium enriched milks. *International Dairy Journal*, *97*, 25–30. https://doi.org/10.1016/j.idairyj.2019.05.001.

Moosavian, S. P., Rahimlou, M., Saneei, P., & Esmaillzadeh, A. (2020). Effects of dairy products consumption on inflammatory biomarkers among adults: A systematic review and meta-analysis of randomized controlled trials. *Nutrition, Metabolism and Cardiovascular Diseases*. https://doi.org/10.1016/j.numecd.2020.01.011.

Mouratidou, T., Vicente-Rodriguez, G., Gracia-Marco, L., Huybrechts, I., Sioen, I., Widhalm, K., Valtueña, J., González-Gross, M., & Moreno, L. A. (2013). Associations of Dietary Calcium, Vitamin D, Milk Intakes, and 25-Hydroxyvitamin D With Bone Mass in Spanish Adolescents: The HELENA Study. *Journal of Clinical Densitometry*, *16*(1), 110–117. https://doi.org/10.1016/j.jocd.2012.07.008.

Murphy, M. M., Douglass, J. S., Johnson, R. K., & Spence, L. A. (2008). Drinking flavored or plain milk is positively associated with nutrient intake and is not associated with adverse effects on weight status in US children and adolescents. *Journal of the American Dietetic Association*, *108*(4), 631–639. https://doi.org/10.1016/j.jada.2008.01.004.

Nurrulhidayah, A. F., Arieff, S. R., Rohman, A., Amin, I., & Shuhaimi, M. (2015). Detection of butter adulteration with lard using differential scanning calorimetry. *Intertiol Food Research Jourl*, *22*(2), 832–839.

Oakley, E., Reinking, J., Sandige, H., Trehan, I., Kennedy, G., Maleta, K., & Manary, M. (2010). A ready-to-use therapeutic food containing 10% milk is less effective than one with 25% milk in the treatment of severely malnourished children. *The Journal of Nutrition*, *140*(12), 2248–2252. https://doi.org/10.3945/jn.110.123828.

Oliveira, D., Reis, F., Deliza, R., Rosenthal, A., Giménez, A., & Ares, G. (2016). Difference thresholds for added sugar in chocolate-flavoured milk: Recommendations for gradual sugar reduction. *Food Research International*, *89*, 448–453. https://doi.org/10.1016/j.foodres.2016.08.019.

Oluk, A. C., Güven, M., & Hayaloglu, A. A. (2014). Proteolysis texture and microstructure of low-fat Tulum cheese affected by exopolysaccharide-producing cultures during ripening. *International Journal of Food Science & Technology*, *49*(2), 435–443. https://doi.org/10.1111/ijfs.12320.

Park, J., Lee, H. S., Lee, C., & Lee, H.-J. (2019). Milk consumption patterns and perceptions in Korean adolescents, adults, and the elderly. *International Dairy Journal*, *95*, 78–85. https://doi.org/10.1016/j.idairyj.2019.03.011.

Qiu, X., Jacobsen, C., & Sørensen, A.-D. M. (2018). The effect of rosemary (Rosmarinus officinalis L.) extract on the oxidative stability of lipids in cow and soy milk enriched with fish oil. *Food Chemistry*, *263*, 119–126. https://doi.org/10.1016/j.foodchem.2018.04.106.

Rodríguez-Pérez, C., Pimentel-Moral, S., & Ochando-Pulido, J. (2019). 4—New Trends and Perspectives in Functional Dairy-Based Beverages. Em A. M. Grumezescu & A. M. Holban (Eds.), *Milk-Based Beverages* (pp. 95–138). Woodhead Publishing. https://doi.org/10.1016/B978-0-12-815504-2.00004-9.

Sani, A. M., Rahbar, M., & Sheikhzadeh, M. (2019). 7 - Traditional Beverages in Different Countries: Milk-Based Beverages. Em A. M. Grumezescu & A. M. Holban (Eds.), *Milk-Based Beverages* (pp. 239–272). Woodhead Publishing. https://doi.org/10.1016/B978-0-12-815504-2.00007-4.

Santaliestra-Pasías, A. M., González-Gil, E. M., Pala, V., Intemann, T., Hebestreit, A., Russo, P., Van Aart, C., Rise, P., Veidebaum, T., Molnar, D., Tornaritis, M., Eiben, G., & Moreno, L. A. (2020). Predictive associations between lifestyle behaviours and dairy consumption: The IDEFICS study. *Nutrition, Metabolism and Cardiovascular Diseases*, *30*(3), 514–522. https://doi.org/10.1016/j.numecd.2019.10.006.

Sonestedt, E., Wirfält, E., Wallström, P., Gullberg, B., Orho-Melander, M., & Hedblad, B. (2011). Dairy products and its association with incidence of cardiovascular disease: The Malmö diet and cancer cohort. *European Journal of Epidemiology*, *26*(8), 609–618. https://doi.org/10.1007/s10654-011-9589-y.

Tamime, A. Y., Hickey, M., & Muir, D. D. (2014). Strained fermented milks – A review of existing legislative provisions, survey of nutritional labelling of commercial products in selected markets and terminology of products in some selected countries. *International Journal of Dairy Technology*, *67*(3), 305–333. https://doi.org/10.1111/1471-0307.12147.

Tanaka, K., Miyake, Y., & Sasaki, S. (2010). Intake of dairy products and the prevalence of dental caries in young children. *Journal of Dentistry*, *38*(7), 579–583. https://doi.org/10.1016/j.jdent.2010.04.009.

Taylan, O., Cebi, N., Tahsin Yilmaz, M., Sagdic, O., & Bakhsh, A. A. (2020). Detection of lard in butter using Raman spectroscopy combined with chemometrics. *Food Chemistry*, *332*, 127344. https://doi.org/10.1016/j.foodchem.2020.127344.

Teegarden, D., & Zemel, M. B. (2003). Dairy Product Components and Weight Regulation: Symposium Overview. *The Journal of Nutrition*, *133*(1), 243S-244S. https://doi.org/10.1093/jn/133.1.243S.

Truong, T., Palmer, M., Bansal, N., & Bhandari, B. (2018). Effects of dissolved carbon dioxide in fat phase of cream on manufacturing and physical properties of butter. *Journal of Food Engineering*, *226*, 9–21. https://doi.org/10.1016/j.jfoodeng.2018.01.012.

Uauy, R., & Valenzuela, A. (2000). Marine oils: The health benefits of n-3 fatty acids. *Nutrition (Burbank, Los Angeles County, Calif.)*, *16*(7–8), 680–684. https://doi.org/10.1016/s0899-9007(00)00326-9.

Wang, J., Wu, T., Fang, X., & Yang, Z. (2019). Manufacture of low-fat Cheddar cheese by exopolysaccharide-producing Lactobacillus plantarum JLK0142 and its functional properties. *Journal of Dairy Science*, *102*(5), 3825–3838. https://doi.org/10.3168/jds.2018-15154.

Wang, X. J., Jiang, C. Q., Zhang, W. S., Zhu, F., Jin, Y. L., Woo, J., Cheng, K. K., Lam, T. H., & Xu, L. (2020). Milk consumption and risk of mortality from all-cause, cardiovascular disease and cancer in older people. *Clinical Nutrition*. https://doi.org/10.1016/j.clnu.2020.03.003.

Weaver, C. M., & Boushey, C. J. (2003). Milk—Good for bones, good for reducing childhood obesity? *Journal of the American Dietetic Association*, *103*(12), 1598–1599. https://doi.org/10.1016/j.jada.2003.10.037.

Xu, P. P., Yang, T. T., Xu, J., Li, L., Cao, W., Gan, Q., Hu, X. Q., Pan, H., Zhao, W. H., & Zhang, Q. (2019). Dairy Consumption and Associations with Nutritional Status of Chinese Children and Adolescents. *Biomedical and Environmental Sciences*, *32*(6), 393–405. https://doi.org/10.3967/bes2019.054.

Yousefi, M., & Jafari, S. M. (2019). Recent advances in application of different hydrocolloids in dairy products to improve their techno-functional properties. *Trends in Food Science & Technology*, *88*, 468–483. https://doi.org/10.1016/j.tifs.2019.04.015.

Yun, B., Maburutse, B. E., Kang, M., Park, M. R., Park, D. J., Kim, Y., & Oh, S. (2020). Short communication: Dietary bovine milk–derived exosomes improve bone health in an osteoporosis-induced mouse model. *Journal of Dairy Science*. https://doi.org/10.3168/jds.2019-17501.

Zittermann, A. (2016). Nutritional and Health-Promoting Properties of Dairy Products: Bone Health☆. Em *Reference Module in Food Science*. Elsevier. https://doi.org/10.1016/B978-0-08-100596-5.21171-2.

Reviewed by:

- *Professor Maria João Barroca* from Agrarian School of Polytechnic Institute of Coimbra, Portugal;
- *Professor João Carlos Gonçalves* from Agrarian School of Polytechnic Institute of Viseu, Portugal.

INDEX

C

D

E

F

G

H

I

J

K

L

M

N

O

P

Q

R

S